LABORATORY MANUAL
Blei and Odian's
General, Organic, and Biochemistry
Connecting Chemistry to Your Life

SARA SELFE

W. H. Freeman and Company
New York

ISBN 0-7167-3582-2 (EAN: 9780716735823)

Copyright © 2000 by W. H. Freeman and Company. All rights reserved.

No part of this book may be reproduced by any mechanical, photographic, or electronic process, or in the form of a phonographic recording, nor may it be stored in a retrieval system, transmitted, or otherwise copied for public or private use, without written permission from the publisher.

Printed in the United States of America

Second printing

Table of Contents

■	Introduction	v
■	Laboratory Safety	1
Experiment 1.	Safety in the Laboratory	5
Experiment 2.	Material Safety Data Sheets (MSDS)	9
Experiment 3.	Error Analysis	15
Experiment 4.	Hot Packs and Cold Packs	23
Experiment 5.	Separation of the Components of a Mixture	31
Experiment 6.	Construction of a Density Column	37
Experiment 7.	Emission Spectra	43
Experiment 8.	Molecular Geometry	51
Experiment 9.	Solubility of Polar and Nonpolar Compounds	57
Experiment 10.	Paper Chromatography	65
Experiment 11.	Stoichiometry and the Chemical Equation	75
Experiment 12.	Charles's Law	83
Experiment 13.	Electrophoresis of Dye Molecules	89
Experiment 14.	Reactions of Common Metals	95
Experiment 15.	Catalysis	101
Experiment 16.	Determination of pH by Using Vegetable Indicators	109
Experiment 17.	Acid-Base Titration	115

Experiment 18.	Simple Distillation of Organic Solvents	121
Experiment 19.	Physical and Chemical Properties of Saturated and Unsaturated Hydrocarbons	127
Experiment 20.	Molecular Shape and Isomerism in Organic Chemistry	135
Experiment 21.	Organic Qualitative Analysis	141
Experiment 22.	Differential Extraction of Organic Acids	149
Experiment 23.	The Synthesis of an Ester	157
Experiment 24.	Synthesis of Aspirin	167
Experiment 25.	Thin Layer Chromatography of Analgesics	175
Experiment 26.	Extraction of Caffeine from Beverages	181
Experiment 27.	Biochemistry Laboratory Techniques	187
Experiment 28.	Separation of Glucose from Starch	195
Experiment 29.	Hydrolysis of Sucrose	203
Experiment 30.	Synthesis of a Soap	209
Experiment 31.	Titration of Amino Acids	215
Experiment 32.	Determination of the Molecular Weight of an Unknown Protein	223
Experiment 33.	Electrophoresis of Normal and Sickle Cell Hemoglobin	233
Experiment 34.	Separation of Albumin from Serum by Affinity Chromatography	239
	Step 1: Affinity Chromatography of Serum	
	Step 2: Analysis of Separation by Electrophoresis	
Experiment 35.	Analysis of Wheat Germ Acid Phosphatase	249
Experiment 36.	Electrophoresis of DNA and DNA Fragments	259
	Part A: Enzyme Digest of DNA	
	Part B: Electrophoresis of DNA Fragments	
Appendix.	Hints on Graphing	265

Introduction

This manual contains a wide variety of experiments that cover the basic topics found in general, organic and biochemistry. You will find that there are more experiments than can usually be accomplished within a one-year course. The number of experiments provides the instructor opportunity to select experiments that best coordinate with their particular course and/or focus. We hope you will agree that the experiments run the spectrum from the classic to the innovative.

Most experiments can easily be completed within a three-hour time frame. Several of the experiments would more aptly be called exercises and can be completed outside the laboratory. But the concepts addressed in these exercises are important to laboratory work or conceptual understanding of chemistry. For example, there is an exercise that acquaints the students with Material Safety Data Sheets (MSDS). As future employees, the students need to be able to read and understand MSDS. This exercise can be introduced at the beginning of the year or just before the organic section. There is also an exercise that covers error analysis. We often ask students to comment on the quality of their data and do not instruct them on the tools we use as chemists to assess quality. This exercise covers accuracy, precision, mean and standard deviation, but also gives students a chance to design an experiment and set their own parameters.

The availability of economically priced electrophoresis equipment has allowed the inclusion of experiments that utilize concepts used in biotechnology. There is an experiment that explores the basis of electrophoresis along with several experiments that utilize electrophoresis to explore biochemical concepts. In addition to electrophoresis, the student will also be introduced to chromatography techniques (affinity and gel filtration) that are used in the modern biochemistry laboratory.

The experiments include data sheets and questions that lead students through the analysis of their data. These sheets do not preclude the assignment of a more formal lab report within the context of the course. The data sheets and questions can serve as an organizational tool for the student.

These laboratory experiments are designed to demonstrate the fundamentals of chemistry and how they are relevant to students' lives and future careers. The study of General, Organic, and Biochemistry can open students' minds to an entirely new perspective on the world that surrounds them. Many thanks to the students who have helped shed light on how best to expose them to this new world. I would also like tho thank Mary Harty and Jim Patterson, without whose help this manual would not exist.

Laboratory Safety

Accidents will occur in the best-regulated families.

Charles Dickens

While Dickens may have felt that his observation was true, the goal of laboratory safety is to prevent accidents in the "best-regulated laboratories." One of the most important facets of a chemistry course is teaching you how to work safely in the laboratory. With a good understanding of common safety rules, the laboratory is a very safe place to work. Your laboratory experience should be a safe one, if you heed the following precautions.

General Safety Rules

Safety goggles must be worn in the laboratory at all times. Wearing goggles is one of the most important things you can do to protect yourself in a laboratory. The use of contact lenses in the laboratory is discouraged and, in some laboratories, prohibited. Check with your instructor.

Learn the location and operation of the safety equipment. These items may include safety showers, emergency eyewashes, fire blankets, and fire extinguishers. If you are not familiar with the operation of a fire extinguisher, ask your instructor to explain it to you. Also, become familiar with all the exits from the laboratory. If a fire alarm goes off while you are in the laboratory, turn off any open flames and electric heaters, grab your valuables, and follow instructions given to you by your lab instructor.

Dress appropriately for the laboratory. Bare feet, sandals, or other open-toed shoes are not a good idea in the laboratory. Cotton clothing (including denim) is particularly susceptible to being eaten by acid solutions. The laboratory is not a good place to wear your favorite clothes. When working with flames, keep long hair tied back. Keep coats, backpacks, and other nonessential materials away from areas where people are working.

Do not eat, drink, or smoke in the laboratory. To avoid contamination, do not bring consumable materials into the lab.

Do not leave a Bunsen burner or other heating apparatus unattended. Turn off open flames if you must leave your area. Bunsen burner flames are often barely visible. Hair may catch on fire or a severe burn may result from leaning or reaching over an invisible flame.

Be aware of others in the laboratory. When working in the laboratory, you must be alert to those around you. Other students will be carrying chemicals to their workstation, so be careful when you walk through the laboratory. Also, if you are working with a flammable chemical, check for open flames in the laboratory.

Working with Chemicals

Follow experimental procedures explicitly, checking and double-checking the identity of all reagents before you use them. There are potentially hazardous combinations of chemicals present in the laboratory. If you have an idea for further investigation, discuss it with your instructor and get authorization. *Never attempt any unauthorized or unassigned experiments.*

Clean up spills immediately. The next person to come along has no way of knowing whether the clear liquid or white powder on the lab bench is innocuous or hazardous. Neutralize acid spills with sodium bicarbonate (baking soda) before cleaning them up. Mercury spills require special procedures; notify the lab assistant if there is a mercury spill.

Wash your hands frequently after handling chemicals and always before leaving the laboratory.

Never return unused reagents to their storage containers. This practice is required to prevent contamination of the whole container. If you take more than you need, dispose of the excess in the appropriate manner. Use the reagents sparingly—they are expensive and time-consuming to prepare. When taking reagents, transfer the amount you need to a labeled clean beaker or other suitable container. Again, check and double-check the identity of all materials before using them.

Do not point the open end of a test tube or other vessel containing a reaction mixture toward yourself or anyone else. If the procedure calls for you to observe the odor of the contents of a vessel, hold it upright 4 to 6 inches front of you, gently fan some of the vapors toward your nose, and sniff gently. This is called **wafting**.

Waft: To carry the vapors to your nose using a cupped hand.

Always add acid to water slowly to dilute. If you attempt to add water to concentrated acid, the heat of solution may vaporize the water and splash concentrated acid in your face. *Sulfuric acid* must be diluted particularly slowly because it releases a tremendous amount of heat upon dilution.

Dispose of reagents and other materials properly. The proper disposal of reagents is essential to the health and safety of school faculty, staff, students, and the surrounding community. Reagents must be managed and discarded in the most responsible and environmentally sound method available. Only specified nonhazardous water-soluble materials can be rinsed down the drain. Waste containers for other materials will be provided. If you are unsure of how to dispose of a particular material, ask your instructor. Dispose of all broken glassware and other sharp objects into the appropriate container. Broken glass placed into the regular trash endangers the custodial personnel.

Wash chemicals off your skin immediately with copious amounts of water. Use eyewash fountains if you get any chemicals in your eye (of course, this accident can be avoided by wearing goggles).

Hazard Identifications

Material Safety Data Sheets (MSDS) are provided by the manufacturer or vendor of a chemical. They contain information about the physical and chemical properties of the chemical and identify any hazards associated with the chemical. They also identify any special handling precautions and protective equipment needed when working with the chemical. You should be familiar with the MSDS before working with any chemical.

Read chemical labels carefully. Chemicals are rated from 0 to 4 according to the hazard they impose, with 0 representing no hazard and 4 representing high hazard. An example of a hazard diamond label is shown below. Each chemical is rated for health, fire, and reactivity. Special warnings are reserved for the fourth diamond.

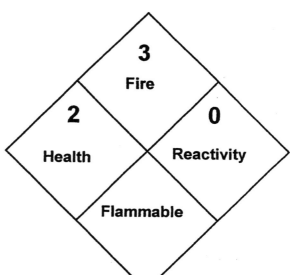

Hazard Diamond colors

Fire = Red
Health = Blue
Reactivity = Yellow

Material Safety Data Sheets (MSDS)

A guide to understanding the contents of a MSDS

Material Safety Data Sheets (MSDS) are provided by the manufacturer or vendor of a chemical. Employers are required to have MSDS available for review by employees. MSDS can be intimidating if you do not understand or are not familiar with the information they contain. In this lesson, we will look at a MSDS and review the information it contains.

A MSDS is divided into several sections. The first section generally contains the name of the compound, any possible synonyms, its chemical formula, and its CAS (Chemical Abstract System) number. The CAS number is used as a method to unambiguously identify a compound. This system is needed because many compounds have one or more common names in addition to their formal chemical names. This lesson contains a sample MSDS for sucrose. Look at the example and you will see that one synonym for sucrose is dextrose; but you probably know it as just plain "sugar."

The next MSDS section contains information about the physical properties of the chemical, such as color, density, boiling point, and solubility. Other sections contain information on reactivity, disposal, conditions to avoid, what to do in case of a spill, and so on.

MSDSs also identify any hazards associated with the chemical and list any special handling precautions and/or protective equipment needed when working with the chemical. Of particular importance is the health hazard, emergency, and first aid data they contain. In the section that includes health hazard data, you will often find the term LD_{50}. This term is used as an attempt to quantify the degree of toxicity of a compound. The term LD_{50} indicates the dose of a chemical that is lethal to 50% of the population. For example, strychnine sulfate, a violent poison, has a LD_{50} of 5 mg/kg when administered orally to rats. That is, if a population of rats is given strychnine sulfate in the amount of 5 mg/kg body weight, 50% of these rats will die.

> The notation used is LD_{50} (ORAL, RAT): 5 mg/kg

Various abbreviations are used for both the species of animal tested and the mode of administration of the compound:

Animal tested
GPG   guinea pig
MUS   mouse

Mode of administration
i.m. or mus.   injected intramuscularly
i.p.   injected intraperitoneally (in the abdominal cavity)
i.v.   intravenously
suc.   subcutaneously

Therefore, a notation of *LD$_{50}$ (i.p., GPG): 200 mg/kg*, would mean that, when a population of guinea pigs was given 200 mg of compound per kg body weight via an injection intraperitoneally, 50% of these guinea pigs died.

The trick in looking at LD_{50}s is to be able to determine whether the compound is a severe poison or not harmful at all. You will see in the following MSDS for sucrose that a LD_{50} is listed. However, we all know that sucrose or, as we call it, table sugar, is not generally considered toxic. Many of the warnings that are used by the manufacturer are stock phrases. For example, the sucrose MSDS suggests that you should induce vomiting if sucrose is ingested, something you probably *shouldn't* do when you use sugar at home. Even though this warning seems ridiculous, you cannot generally ignore the warnings given.

Another important section of a MSDS is the listing of acute and chronic effects. Acute effects are those seen after a short exposure. As a student who may only be exposed to a particular chemical once, pay close attention to acute effects. Chronic effects are due to long-term exposure to a chemical. If you are an employee who works with chemicals, you will also need to be aware of the effects of exposure over time.

One of the goals of this lesson is to give you a sense of what these warnings and numbers mean and to help you gain a level of comfort when working with chemicals in the laboratory. One of the major purposes of taking a chemistry laboratory is for you to learn how to handle chemicals safely. Learning how to read a MSDS is a step toward that goal.

To complete this lesson, you must use MSDSs to determine specific information about sucrose (MSDS provided) and another compound. Several different compounds are listed as possibilities, or you may choose another compound you find interesting.

Choose one of the following:

acetic acid (found in vinegar)
acetylsalicylic acid (aspirin)
formaldehyde
hydrogen peroxide
iodine (used as an antiseptic)
isopropyl alcohol (rubbing alcohol)

menthol
methanol (wood alcohol)
naphthalene (mothballs)
sodium benzoate (a preservative)
sodium bicarbonate
sodium chloride (table salt)

Finding an MSDS on the Web

MSDSs are easy to find on the Web. Using a browser such as Netscape or Internet Explorer, search for "MSDS." This search should give you a list of several databases. The easiest database to use is one with a searchable index.

MSDS for sucrose (abbreviated)

```
=====================================================================
                    Ingredients/Identity Information
=====================================================================
Ingredient: SUCROSE
Synonyms: CANE SUGAR, BEET SUGAR, DEXTROSE
Percent: 100.0
NIOSH (RTECS) Number: WN6500000
CAS Number: 57-50-1
=====================================================================
                    Physical/Chemical Characteristics
=====================================================================
Appearance And Odor: WHITE, ODORLESS CRYSTALS
Boiling Point: NOT GIVEN
Melting Point: DECOMPOSES
Vapor Pressure (MM Hg/70 F): NOT GIVEN
Vapor Density (Air=1): NOT GIVEN
Specific Gravity: 1.59
Decomposition Temperature: 220F,104C
Solubility In Water: FREELY SOLUBLE
=====================================================================
                    Fire and Explosion Hazard Data
=====================================================================
Flash Point: NONE
Lower Explosive Limit: NOT GIVEN
Upper Explosive Limit: NOT GIVEN
Extinguishing Media: DRY CHEMICAL, CARBON DIOXIDE, ALCOHOL FOAM
Special Fire Fighting Procedures: WEAR FULL PROTECTIVE CLOTHING
AND NIOSH-APPROVED SELF-CONTAINED BREATHING APPARATUS WITH FULL
FACEPIECE OPERATED IN THE POSITIVE PRESSURE MODE.
Unusual Fire And Explosion Hazards: AS WITH ANY FINELY DIVIDED
ORGANIC SOLID, DUST MAY BE EXPLOSIVE IF MIXED WITH AIR IN
CRITICAL PROPORTIONS AND IN THE PRESENCE OF AN IGNITION SOURCE.
=====================================================================
                            Reactivity Data
=====================================================================
Stability: YES
Conditions To Avoid (Stability): EXTREME HEAT
Materials To Avoid: NITRIC ACID, SULFURIC ACID
Hazardous Decomp Products: CARBON MONOXIDE, CARBON DIOXIDE MAY
BE FORMED.
```

Health Hazard Data

LD50-LC50 Mixture: LD50 (ORAL, RAT) IS 29700 MG/KG.
Route Of Entry - Inhalation: YES
Route Of Entry - Skin: YES
Route Of Entry - Ingestion: YES
Health Hazard Acute And Chronic: EYE CONTACT MAY CAUSE IRRITATION. SKIN CONTACT MAY CAUSE IRRITATION. INGESTION MAY CAUSE GASTROINTESTINAL DISCOMFORT. INHALATION MAY CAUSE IRRITATION TO RESPIRATORY TRACT.
Carcinogenicity - NTP: NO
Carcinogenicity - IARC: NO
Carcinogenicity - OSHA: NO
Signs/Symptoms Of Overexp: EYE IRRITATION, SKIN IRRITATION, GI TRACT IRRITATION, RESPIRATORY TRACT IRRITATION.
Emergency/First Aid Proc: EYES: FLUSH WITH WATER FOR 15 MINUTES WHILE HOLDING EYELIDS OPEN. GET MEDICAL ATTENTION. SKIN: REMOVE CONTAMINATED CLOTHING. WASH WITH SOAP AND WATER. IF IRRITATION PERSISTS, GET MEDICAL ATTENTION. INHALATION: REMOVE TO FRESH AIR. RESTORE BREATHING. GET MEDICAL ATTENTION. INGESTION: GIVE WATER AND INDUCE VOMITING AS DIRECTED BY MEDICAL PERSONNEL. IMMEDIATELY CALL A PHYSICIAN OR POISON CONTROL.

Precautions for Safe Handling and Use

Steps If Matl Released/Spill: VENTILATE AREA OF SPILL. CLEAN-UP PERSONNEL SHOULD WEAR PROPER PROTECTIVE EQUIPMENT. AVOID CREATING DUST. SWEEP OR SCOOP UP AND CONTAINERIZE FOR DISPOSAL.
Neutralizing Agent: NONE SPECIFIED BY MANUFACTURER.
Waste Disposal Method: WHATEVER CANNOT BE SAVED FOR RECOVERY MAY BE BURNED IN AN APPROVED INCINERATOR OR DISPOSED IN AN APPROVED WASTE FACILITY. ENSURE COMPLIANCE WITH LOCAL, STATE AND FEDERAL REGULATIONS.
Precautions-Handling/Storing: STORE TIGHTLY CLOSED IN A COOL, DRY, WELL VENTILATED AREA. SUITABLE FOR STORAGE IN ANY GENERAL CHEMICAL STORAGE AREA. DO NOT BREATHE DUST.
Other Precautions: DO NOT GET IN EYES. AVOID PROLONGED OR REPEATED SKIN CONTACT.

Control Measures

Respiratory Protection: NONE NORMALLY REQUIRED. IN DUSTY SITUATION, A NIOSH-APPROVED RESPIRATOR FOR DUST MAY BE WORN.
Ventilation: LOCAL EXHAUST AND MECHANICAL (GENERAL) VENTILATION AS REQUIRED TO MAINTAIN EXPOSURE LEVELS.
Protective Gloves: RUBBER, NEOPRENE, PVC OR EQUIVALENT
Eye Protection: SPLASH PROOF CHEMICAL SAFETY GOGGLES
Other Protective Equipment: CHEMICAL RESISTANT CLOTHING AS NECESSARY TO PREVENT SKIN CONTACT. AN EMERGENCY EYEWASH AND SHOWER SHOULD BE AVAILABLE.
Work Hygienic Practices: WASH HANDS THOROUGHLY WITH SOAP AND WATER BEFORE EATING, DRINKING, SMOKING OR USING TOILET FACILITIES.

Experiment 3

Error Analysis

An expert is a man who knows some of the worst errors that can be made in the subject in question and who therefore understands how to avoid them.

Werner Heisenberg, Nobel Laureate 1932

Heisenberg is famous in chemistry for his statement of the Heisenberg Uncertainty Principle, which asserts that you cannot know the position and momentum of a subatomic particle simultaneously. This constraint is a result of the fact that in the process of making a measurement of one factor you affect the other. You might say that Heisenberg was an expert in error. Uncertainty does not exist only at the subatomic level. Whenever any measurement is made, some uncertainty exists. As the quote above states, if you want to be an expert, or a good scientist, it is important to know where uncertainties and possibilities of error lie.

In any scientific experiment, it is necessary for you to analyze your results and determine their dependability. We evaluate our results by evaluating the error in them. The word *error* in this case does not imply that one has made a mistake; rather it means that, even with the best technique, measured values are never infinitely accurate but contain some uncertainty. Error analysis is the process of evaluating the experimental procedure and the results of that procedure and determining where uncertainties arise. It is an important aspect of experimental design. It is only when you know where things can go wrong that you can "fix them," or at least reduce the probability of error.

Sources of Error

All experiments are subject to the possibility of error. There are two types of error. First, there is **systematic error**. For example, in taking a measurement, the instrument you are using may be erroneous. If the scale you use to weigh as mass is not calibrated correctly, every time you weigh something the result will be wrong. For this type of error, repetition of the measurement will not improve things, because the error is repeated systematically every time. This error can be avoided if the instrument you are using is calibrated before you do the experiment.

The second type of error is **random error.** Random error can result from a large number of unpredictable sources, for example, poor eyesight, line voltage variation, vibrations from a passing truck, tiredness. However, the important aspect of random error is that it is random and hard to predict. The way to avoid this error is to take many measurements. One random error in many measurements will have minimal impact.

Accuracy and Precision

In an attempt to minimize error, we try to be accurate and precise. Accuracy and precision are two different concepts. A result is **accurate** when you are close to the true value. A result is **precise** when each measurement in a series agrees with the others. The important point is that you can have a series of precise measurements that are not close to the true value (precise measurements need not be very accurate). A systematic error can lead to a series of measurements that are precise but not accurate.

ACCURATE AND PRECISE

A useful value in calculating accuracy and precision is the mean, or average, of all the measurements, represented by $\bar{x}$. It is calculated as the sum of all the measurements divided by the number of measurements:

PRECISE BUT NOT ACCURATE

$$\bar{x} = \frac{x_1 + x_2 + x_3 + x_4 + \ldots + x_n}{n}$$

Accuracy refers to the difference between your measured value and the true value. The ultimate goal of any analysis is to have the average measured value be the same as the true value.

By calculating a percent (%) deviation of the measured value from the true value, we get a mathematical value for accuracy of the measurement. The % deviation is calculated as

$$\% \text{ deviation} = \frac{|\bar{x} - x_{true}|}{x_{true}} \cdot 100$$

where $\bar{x}$ is the mean of the measured values and x_{true} is the true value.

Precision indicates the consistency in a set of measurements. Precision can be influenced by how consistent you are in performing the technique. For example, with experience, the precision of pipetting a volume of a liquid can be very good.

There are several ways to express the degree of precision in your measurements. The measure of precision you are probably most familiar with is the use of the standard deviation. Often results from exams are reported as the mean and the standard deviation.

The **standard deviation** measures the size of the average error (or average deviation) in a set of repeated measurements. It is computed by taking the difference between the mean, $\overline{x}$, and each individual measurement, x_i, squaring this value, adding all these numbers together, dividing by the number of measurements minus one, $(n-1)$, and then taking the square root.

$$s = \sqrt{\frac{(x_1-\overline{x})^2 + (x_2-\overline{x})^2 + \cdots + (x_n-\overline{x})^2}{n-1}}$$

The value of the standard deviation is commonly used to indicate the range of experimental values. What does this tell us? When the mean and standard deviation are calculated they give you the range into which 68% of the experimental values will fall. Now when a professor tells you that the mean for a test was 85% ± 15, you will know that 68% of the grades for the exam fell between 100 and 70.

In experiments where you take multiple measurements, you will be able to determine the mean and standard deviation. As you increase the number of measurements, generally the deviation from the mean decreases. The effect of random error is minimized. For this reason, it is necessary to take multiple measurements in an experiment. Although it is difficult to single-handedly take hundreds of measurements, you can often get good results by combining the measurements of a whole class. In this exercise, you will have an opportunity to decide how to take your own measurements. In addition, you will get a chance to practice calculating a mean and the standard deviation.

NAME _____

Experimental Design

Working with a group of students, you will design an experiment to determine the number of chocolate chips in a chocolate cookie.

Names — List the names of all students in your group.

Hypothesis — What is your prediction for the number of chips in a cookie?

Purpose (given) — Determine the number of chips in a representative chocolate chip cookie.

Procedure — Describe the technique your group is going to use to count the number of chips.

Data — For each cookie, record the number of chips.

Cookie #	Number of chips
_____	_____
_____	_____
_____	_____
_____	_____
_____	_____
_____	_____
_____	_____
_____	_____
_____	_____

Brand of cookie: _____

Results   Calculate the average number of chips in a cookie.

Conclusion   How good is your result? (**Hint:** What is the standard deviation?)

QUESTIONS

1. Do you think your results represent a reasonable value for the number of chips in a cookie?

2. How could you revise the experiment to increase the accuracy of your results?

3. What factors do you think would cause your results to differ from the results of another group of students?

Experiment 4

Hot Packs and Cold Packs

Heat, temperature and specific heat

Depending on the circumstances, a doctor or nurse may prescribe the use of hot or cold treatments to aid in the relief of pain. The application of heat will increase circulation and help reduce inflammation. By contrast, cold is used to reduce swelling and decrease pain. Commercially available hot packs and cold packs utilize the heat given off or absorbed when compounds dissolve or crystallize. In this experiment, you will observe the temperature changes that occur when calcium chloride and ammonium chloride are dissolved in water. From these data, you will be able to calculate the heat given off or absorbed during the dissolution.

> Heat is measured in °C (Celsius), °F (Fahrenheit), or K (Kelvin).

Heat is not the same as temperature. **Temperature** does not depend on the quantity of the material; it is simply a measure of how hot or cold a sample is. **Heat**, on the other hand, does depend on the quantity of the material and is a measure of how much energy a sample contains. For example, compare the temperature and the heat content of a cup of freshly brewed coffee and an urn of freshly brewed coffee. The coffee in each case will be approximately the same temperature, but an urn full of coffee contains more heat than a cup of coffee simply because there is more coffee.

Heat content is dependent not only on the quantity of a substance but also on the identity of the substance. Each substance has its own specific heat. The specific heat of water is 1.0 calorie/g°C. This statement means it takes 1.0 calorie of energy to raise the temperature of 1.0 g of water by 1 degree Celsius. Ethyl alcohol has a specific heat of only 0.59 calories/g°C. Much less energy is needed to raise the temperature of 1 g of ethyl alcohol by 1°C.

> **Specific heat** is defined as the amount of heat needed to raise the temperature of 1 g of the substance by 1°C.

To calculate the heat given off or absorbed during a dissolution reaction, there is a third aspect to consider (besides the mass and the specific heat of the material): temperature change (ΔT). The calculation is expressed as

Heat given off or absorbed = (specific heat)(mass)(temperature change)

The particular reactions in this experiment are those that take place in commercially available hot and cold packs. The packs consist of an outer pouch containing the solid and an inner pouch containing water. When the inner pouch is broken, the dissolution takes place with an accompanying release or absorption of heat. Some hot packs are based on the heat given off when a saturated solution crystallizes.

A process that gives off heat is called **exothermic**, and a process that absorbs heat is called **endothermic**.

MATERIALS

Equipment

- thermometer
- stirring rod
- 4 Styrofoam cups
- 100-mL graduated cylinder
- spatula

Chemicals

- calcium chloride, $CaCl_2$
- ammonium chloride, NH_4Cl

SPECIAL NOTES

This experiment is quantitative, so be careful when weighing. You do not need to weigh out exactly 20 g, but you do need to record the mass of the compound after weighing.

WASTE DISPOSAL
The calcium chloride and ammonium chloride are nonhazardous and can be rinsed down the drain with copious amounts of water.

PROCEDURE

1. Using the two Styrofoam cups, construct a simple calorimeter by nesting the two cups, one inside the other.

2. Add 100 mL of water to the calorimeter. Allow the water to stand for several minutes to reach a stable temperature. Record the temperature.

3. Weigh out approximately 20 g of $CaCl_2$. Be sure to record the exact mass.

4. While your partner holds the cup and the thermometer steady, add all the $CaCl_2$ to the calorimeter and stir rapidly with a stirring rod. Be careful not to hit the thermometer while stirring.

5. After mixing, time-temperature points should be recorded in the data table. One partner should read the temperature while the other reads the time and keeps the record. It is best to take temperature readings at frequent intervals, every minute or less, while the temperature is changing dramatically. Take readings at longer

A calorimeter is an apparatus used to measure heat.

CAUTION: Do not use the thermometer as a stirring rod.

intervals, every 2-3 minutes, as the temperature change starts to slow. You will make a plot on graph paper to find the maximum temperature. The maximum temperature reached is the final temperature (T_f).

6. After recording your data, wash the contents of the cup down the drain with lots of water. Rinse off the thermometer and stirring rod.

7. Repeat the above procedure, using approximately 20 g ammonium chloride (NH_4Cl). Be sure to record the exact mass. The dissolution of ammonium chloride is endothermic, so in this case the minimum temperature reached is final temperature (T_f).

Experiment 5

Separation of the Components of a Mixture

Differences in solubility and boiling points will be used to separate four different components.

When substances that do not react chemically are combined, a mixture results. **Mixtures** are characterized by two fundamental properties: first, each of the substances in the mixture retains its chemical integrity, and second, mixtures are separable into these components by *physical* means. In this experiment, you will separate the components of a sand, iodine, and copper sulfate mixture. This particular mixture is not uniform in composition and is called a **heterogeneous mixture**. We encounter heterogeneous mixtures, such as cement and soil, everyday. In contrast, a **homogeneous mixture** has the same composition throughout. Examples of homogeneous mixtures are soft drinks and salt water.

The following techniques will be used to separate the components of the sand, iodine, and copper sulfate mixture.

Sublimation: To change directly from the solid state to the gaseous state without becoming liquid. Unlike most substances, iodine sublimes near room temperature and pressure, and can be sublimed from the mixture.

Extraction: To separate two compounds by taking advantage of differences in solubility. If one substance (solute) is very soluble in a given liquid (solvent) and the other substance is almost insoluble in the same liquid, the first substance may be almost completely separated from the second by merely shaking the mixture with the given solvent. The soluble material goes into solution, whereas the insoluble substance remains behind. In this experiment, copper sulfate will be extracted into (will be dissolved in) water and the sand will remain undissolved.

Filtration: To separate a solid from a liquid by means of a porous barrier that allows the liquid to pass through but not the solid. You will use this method to separate the solution of copper sulfate from the insoluble sand.

Evaporation: To separate the components of a mixture by taking advantage of the differences in boiling points. In this experiment, you will heat the copper sulfate solution until the water has evaporated and only the copper sulfate remains behind.

The following flow diagram summarizes the separation:

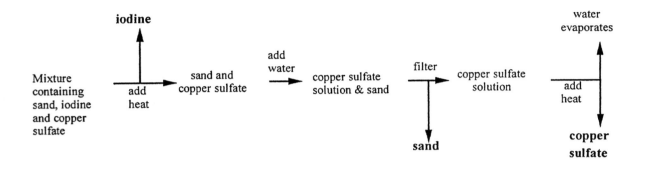

MATERIALS

Equipment

- ✓ Ring stand with ring
- ✓ ~~Bunsen burner~~
- ✓ filter paper
- ✓ wire gauze
- ✓ 150-mL beaker
- ✓ evaporating dish or watchglass
- ✓ funnel
- ✓ 100-mL beaker
- ✓ 100-mL graduated cylinder

Chemicals

- ✓ mixture of sand, copper sulfate, and iodine

PROCEDURE

Part A: Sublimation of iodine

- Clamp an iron ring stand at a height sufficient to allow a Bunsen burner to be placed beneath it and then place a square of wire gauze on the ring.

Experiment 6

Construction of a Density Column

Exploration of the physical property density

In this experiment, you will be constructing a column composed of four distinct liquid layers. The separation of these layers is due to two different physical properties—solubility and density. One solution is soluble in another if, when they are mixed, they form a homogeneous mixture. For example, if you have done any cooking, you realize that oil is not soluble in water. You can mix oil and water by shaking but in time, they will separate. By contrast, ethanol is soluble in water; they mix completely and do not separate upon standing. The old saying "like dissolves like," summarizes this property. Compounds that are similar in their interactions with each other, in most cases will be soluble. Thus, polar materials are miscible with other polar materials and non-polar materials are miscible with other non-polar materials. However, polar and non-polar materials are not soluble with each other.

> A homogeneous mixture is a mixture in which the composition is the same throughout.

Density is a physical property and is defined as, the mass of a material per given unit of volume. It is probably a familiar concept; we know a metal weight will sink in water because it weighs more than the same volume of water. To determine the density of an object you need two pieces of information: the volume and the mass of that volume. Density of a liquid is determined by using the expression

$$\text{Density} = \frac{\text{mass in grams}}{\text{volume in mL}}$$

The density of a liquid is given in grams per milliliter (g/mL) because the milliliter is the unit commonly used to measure liquids. A different unit is used for solids. The volume of a solid, if it is a cube, can be determined by multiplying the height (in centimeters, cm), the width (in cm), and the length (in cm), to get cm^3. Therefore, the mass of a solid is given in grams per cm^3.

The four liquids you will be placing in your density column are water, glycerol, saturated potassium bromide solution, and hexane— each with different densities and polarities. The potassium bromide solution and water are strictly polar and hexane is strictly nonpolar, but the glycerol has a molecular structure that results in a polarity that is intermediate. In this experiment, you will determine the density differences of these liquids in order to create a density column. You will then drop different solids into your density column. Blocks will settle below a layer of a lesser density and on top of a layer of greater density. You will then use this information to determine the approximate densities of different blocks of solids.

MATERIALS

Equipment

- ✓ 10-mL graduated cylinder(s)
- ✓ 100-mL graduated cylinder
- ✓ 100-mL beaker
- ✓ blocks made from different materials

Chemicals

- ✓ glycerol, $C_3H_8O_3$
- ✓ hexane, C_6H_{14}
- ✓ saturated potassium bromide solution, KBr in H_2O
- ✓ water, H_2O

SAFETY CONSIDERATIONS

Caution: Hexane is flammable, so keep it away from open flames. Dispense hexane in the hood. Hexane should not be disposed of down the drain.

PROCEDURE

Part A: Determination of the density of the liquids

1. Weigh a 10-mL graduated cylinder. Record the mass in the *Data* Section.

2. Add *exactly* 5.0 mL of one of the liquids to the graduated cylinder.

3. Weigh the graduated cylinder plus liquid. Record the mass in the *Data* Section.

4. In the *Results* section, calculate the density of the liquid by using the following equation:

$$\text{Density} = \frac{\text{mass of graduated cylinder with liquid - mass of graduated cylinder empty}}{5 \text{ mL}}$$

5. Put the first liquid aside in a container. Repeat Steps 1–4 for each liquid. If you must use the same graduated cylinder, be sure to rinse out all traces of the previous liquid before adding a new liquid. After weighing, place each liquid in a separate container until ready to make the density column.

Part B: Construction of the density column

After you have calculated the densities of the individual solutions you can begin the construction of the density column. Verify the order before beginning.

1. Carefully pour the 10 mL of the densest liquid into a 100-mL graduated cylinder.

2. Carefully pour the 10 mL of the next densest liquid into the 100-mL graduated cylinder. Pour gently to minimize mixing of layers.

3. Carefully follow this with the next densest liquid.

4. Finally, add the least dense liquid, being careful not to allow the liquids to mix. If you have poured the liquids carefully, you should have a four-layer column at this point. Record your observations.

5. Determine the approximate density of two different materials supplied in the laboratory by gently dropping them into the liquid column. You may need to apply a slight pressure (poke) to the materials to break any surface tension that exists. Record between which two layers the material rests.

6. Mix the layers in the column by pouring the contents of the column back and forth five times between the graduated cylinder and a 100-mL beaker. Record your observations.

Experiment 7

Emission Spectra

Just like astronomers, you will identify elements by using their emission spectra.

When sunlight passes through a raindrop, a rainbow of colors called a **continuous spectrum** is produced. Likewise, when white light from an incandescent bulb is passed through a prism, it is separated into a rainbow of colors. However, when light from a mercury lamp is passed through a prism, we do not see all the colors of a continuous spectrum. Instead, several discrete colored bands of light called **spectral lines** are observed. When an atom absorbs energy from a flame or electrical discharge, its electrons are excited to a higher energy state. When the electron returns to its original energy state, it emits the energy previously absorbed in the form of one or more photons (packages of light). The photon energy, E_{photon}, is related to the wavelength and frequency of the light by the following equation:

$$E_{photon} = h(c/\lambda) = h\nu$$

where c is the speed of light traveling through a vacuum (3×10^8 m/s) and h is Planck's constant (6.63×10^{-34} J·s/photon). The wavelength (λ) of the photon is in meters and is inversely proportional to the frequency (ν) of the photon in s^{-1} (or hertz, Hz).

Each element has its own characteristic set of spectral lines. We can use this information to identify elements. For example, mercury has three very distinct lines in its visible spectrum: a violet line with a wavelength of 436 nm, a green line with a wavelength of 546 nm, and a yellow line with a wavelength of 580 nm. Additionally, the above equation shows us the relationship of the wavelength to the energy of the photon. The equation states that the energy of the photon is inversely proportional to the wavelength. That is, the longer (larger) the wavelength, the lower the energy of the photon.

In the laboratory, you will use a spectroscope to see the spectral lines of mercury. The spectroscope consists of a box with a slit and ruler at one end and a diffraction grating at the other end (the diffraction grating is our stand-in for a prism). The light will enter through the slit, and the grating will diffract the light into its individual lines. The lines will project onto the ruler. The spectroscope has not been calibrated; that will be your first task. To calibrate, you will observe where each spectral lines falls on the ruler and plot the ruler marking versus the wavelength of the lines. Once the spectroscope is calibrated, you will look at several other lamps and flames to determine their characteristic line spectra. Finally, you will be asked to identify an unknown sample by its line spectrum.

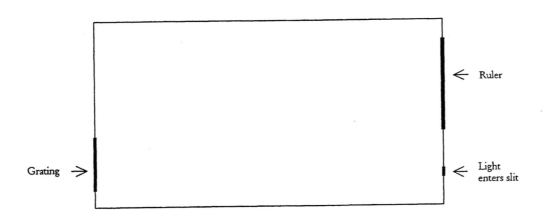

MATERIALS

Equipment

- ✓ spectroscope
- ✓ power supplies with discharge tubes
- ✓ cotton swabs

Chemicals

- ✓ sodium chloride (NaCl)
- ✓ potassium chloride (KCl)
- ✓ lithium chloride (LiCl)
- ✓ strontium chloride ($SrCl_2$)
- ✓ barium chloride ($BaCl_2$)

PROCEDURE

Part A: Calibration of the Spectroscope

All scientific instruments must be calibrated before any quantitative measurements can be made. The room will need to be darkened so that you can calibrate your spectroscope. Turn on the power supply containing the mercury discharge tube. The tube will glow with the characteristic color (pink hue) of a mercury discharge lamp. Look through the grating of the spectroscope at the discharge tube to observe the line spectrum. The spectrum that you see will consist of three prominent lines of light. Match each spectral line to a line on the ruler inside the spectroscope. The yellow line is 580 nm, the green line is 546 nm, and the violet line is 436 nm. Once you have determined the ruler mark for each of these lines, you will need to plot a graph of "ruler mark" versus "wavelength." This graph is the spectroscope calibration line that will enable you to determine the wavelengths of unknown spectral lines. There may be some other faint lines. Identify those that you can see.

Part B: Helium spectrum

Now look through the spectroscope at a helium lamp. Write down the colors of any lines you observe. Again, match each spectral line to a line on the ruler. Use the calibration graph to determine the wavelength of each line.

Part C: Strontium, lithium, potassium, sodium, and barium spectra

CAUTION! Be careful not to get too close to the flame.

In the laboratory, you will find the salts of each of the above elements. Have your partner dip a cotton swab in the aqueous saturated solution of one of the salts. (For the best results, try to get crystals of the salt onto the cotton swab.) Light a Bunsen burner and have your partner hold the cotton swab in the flame. Each salt will impart a distinct color to the flame. Look through the spectroscope at the flame. You will need to be close to the flame; your partner must make sure your do not set the spectroscope on fire! Write down the color and ruler marking for each line. Use the calibration graph to determine the wavelength. Repeat this process for each salt, using a clean cotton swab each time to avoid contamination of the salts.

Part D: Identification of unknown

In the laboratory, you will find unknown salts. Determine the identity of one of the unknown salts by observing its line spectra. The unknown will be one of the above salts.

NAME _____

DATA AND RESULTS

Mercury Emission Spectrum

Spectral line	Ruler mark	Wavelength (nm)
yellow line		580 nm
green line		546 nm
violet line		436 nm

Other observations.

Graph of wavelength versus ruler mark

You will use this graph to determine the wavelengths of the spectral lines of the other elements tested.

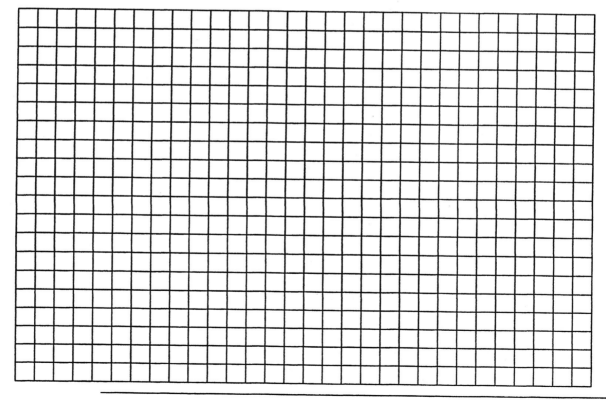

Helium Emission Spectrum

Spectral line	Ruler mark	Wavelength (nm)

Strontium Emission Spectrum

Spectral line	Ruler mark	Wavelength (nm)

Lithium Emission Spectrum

Spectral line	Ruler mark	Wavelength (nm)

NAME _____

Potassium Emission Spectrum

Spectral line	Ruler mark	Wavelength (nm)

Sodium Emission Spectrum

Spectral line	Ruler mark	Wavelength (nm)

Barium Emission Spectrum

Spectral line	Ruler mark	Wavelength (nm)

Identification of Unknown

Note your observations and use the spectral lines to identify your unknown.

Unknown identity _____

QUESTIONS

1. Sodium emits a very simple line spectrum consisting of two orange lines with wavelengths of 589 nm and 591 nm. What are the frequencies of these two spectral lines? What is the energy of these two spectral lines?

2. Which has more energy, green light or orange light? Explain your choice.

Experiment 8

Molecular Geometry

The shape of a covalent molecule gives us insight into its properties.

Molecules are three-dimensional objects. Paper and projection screens are two-dimensional. We are used to looking at photographs of people, and in our minds we can take that two-dimensional photograph and imagine the three-dimensional face. We have no frame of reference for "seeing" a molecule in three dimensions. This simple fact-of-life means that a student of chemistry seldom gets to "see" molecules as they actually are and consequently often have difficulty understanding why they behave as they do. In this exercise, you will construct three-dimensional models of a number of simple molecules and then examine these models to see what molecular properties is a consequence of their shape.

You will use a molecular model kit to construct the molecules as they are discussed in this exercise. For each model, you will first draw a Lewis dot structure, including nonbonding electrons. The **Lewis dot structure** is a two-dimensional representation that shows the arrangement of atoms in a molecule. The Lewis dot structure includes both bonding and nonbonding electrons. When drawing covalent molecules, remember that the electrons are shared between two atoms, forming a covalent bond.

Lewis Structures

The following steps will help you draw Lewis structures.

Note: Remember the octet rule! Atoms tend to form bonds so that their electron configuration is the same as that of the noble gas nearest to them in the periodic table. This configuration would have eight electrons for the first three rows in the periodic table, except for hydrogen, which forms only one bond and has only two electrons.

Electronegativity describes the ability of an atom covalently bonded to another atom to draw the bonding electrons toward itself.

1. Polyatomic (more than two atoms) molecules and ions often consist of a central atom surrounded by more **electronegative** atoms (hydrogen is the exception). You might say that the molecule tries to be as symmetrical as possible. For example, in a molecule such as carbon tetrachloride, CCl_4, you find the carbon in the center and the chlorines around it.

2. Write the skeletal structure and connect the central atom to the surrounding atoms with a straight line (the bond). Each bond represents two electrons.

3. Determine the total number of valence electrons in all the atoms in the molecule. For a polyatomic ion, add or subtract electrons to arrive at the appropriate charge. For example, if the charge is –1, you need to add an electron.

4. Place electrons about the outer atoms so that each atom (except hydrogen) has an octet.

5. Subtract the number of electrons assigned so far (include two electrons for each bond) from the total calculated in Step 3. Any electrons that remain are assigned in pairs to the central atom.

6. If a central atom has fewer than eight electrons after Step 5, a multiple bond is likely. Move one or more nonbonding pairs from an outer atom to the space between the atoms to form a double bond.

Let's do an example: the Lewis dot structure of carbon dioxide (CO_2).

Step 1 Choose the central atom. Carbon is less electronegative than oxygen, so carbon is in the center and the oxygens are arranged around it.

Step 2 Write the skeletal structure.

$$O—C—O$$

Step 3 Determine the total number of valence electrons.

4 for carbon	= 4
6 for each oxygen (2 × 6)	= 12
Total	= 16

Step 4 Place electrons.

We have used two electrons in each bond, so there are 12 electrons left and we place those around the outer atoms.

$$:\ddot{O}—C—\ddot{O}:$$

Step 5 Now that we have placed all the electrons around the two oxygen atoms, we discover that the carbon does not have a full octet.

Step 6 For carbon to have a full octet, we need to move electrons between the carbon and the oxygens to form multiple bonds.

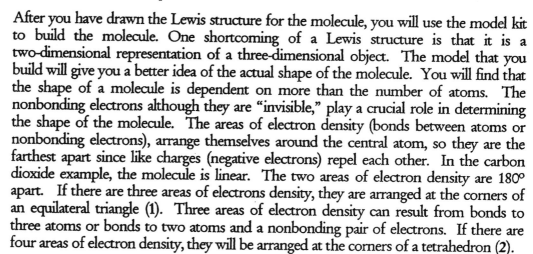

Molecular Shape

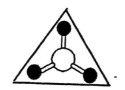

1 Trigonal planar molecule inside triangle

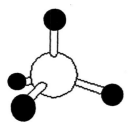

2 Shape of tetrahedral molecule

After you have drawn the Lewis structure for the molecule, you will use the model kit to build the molecule. One shortcoming of a Lewis structure is that it is a two-dimensional representation of a three-dimensional object. The model that you build will give you a better idea of the actual shape of the molecule. You will find that the shape of a molecule is dependent on more than the number of atoms. The nonbonding electrons although they are "invisible," play a crucial role in determining the shape of the molecule. The areas of electron density (bonds between atoms or nonbonding electrons), arrange themselves around the central atom, so they are the farthest apart since like charges (negative electrons) repel each other. In the carbon dioxide example, the molecule is linear. The two areas of electron density are 180° apart. If there are three areas of electrons density, they are arranged at the corners of an equilateral triangle (1). Three areas of electron density can result from bonds to three atoms or bonds to two atoms and a nonbonding pair of electrons. If there are four areas of electron density, they will be arranged at the corners of a tetrahedron (2).

When predicting the shape of the molecule, you need to consider the areas of electron density, as described above. However, when describing the shape of the molecule you only consider the atoms themselves. For example, you will discover that water has four areas of electron density positioned tetrahedrally around the oxygen. But water only has three atoms, so the description of the molecule's shape is bent.

Polarity of Molecules

A characteristic that plays an important role in the behavior of molecules is the distribution of electrons. When the electrons are evenly distributed between two atoms, as in fluorine, F_2, the bond is called nonpolar. When two *dis*similar atoms are bonded, the electrons migrate toward the atom with the higher electronegativity (attraction for electrons), and the bond is polar. In HCl, for example, the stronger effective nuclear charge from the chlorine nucleus pulls the electrons in the bond closer to the chlorine nucleus and farther from the hydrogen nucleus. The net effect is a slightly more negative charge near the chlorine, which is balanced by a slightly more positive charge near the hydrogen.

The little "extra" charges are indicated by writing δ^- near the electronegative atoms and δ^+ near the electropositive ones. This type of covalent bond is called a **polar covalent bond**. Remember that, overall, these molecules are neutral. All covalent bonds between two different elements behave this way; the extent of the polarity depends on the two elements involved. Molecules that contain the highly polar bonds between elements with a large electronegative difference will have a greater separation of positive and negative charge. Of particular interest are the three most electronegative elements: oxygen, nitrogen, and fluorine. For a molecule with more than two atoms, the polarity of each bond and the arrangement of the bonds needs to be considered when predicting the overall polarity of the molecule.

H-Cl
δ^+ δ^-

Because there are two atoms involved in the polar covalent bond, the phrase *dipole moment* is also used (dipole means that there are two poles, or charges: δ^- and δ^+). Almost every bond in a molecule has some polarity, which is often indicated by drawing a little arrow next to the bond, pointing from the electropositive atom to the electronegative atom. (The + sign is created by drawing a line through the tail of the arrow: +—>). The overall effect of these individual polar bonds depends on the symmetry of the molecule. If the molecule has sufficient symmetry, the individual polar bonds cancel one another, leaving a molecule that appears to have a uniform charge distribution. If, on the other hand, the molecule has some sort of irregular shape or if the molecule is not symmetrical, the polar bonds will not cancel, leaving the molecule with a net, or permanent, dipole moment. In the carbon dioxide example, there is a dipole for each carbon-oxygen bond, but they pull in equal and opposite directions and cancel out. The net result is that carbon dioxide is nonpolar.

H-Cl
+—>

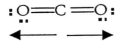   The individual dipoles cancel each other out.

It is important to realize that a polar molecule is a molecule that is neutral but that has an unequal distribution of electrons such that one end of the molecule is positive and the other negative. A molecule that has a charge (a polyatomic ion), such as CO_3^{2-}, is said to be ionic rather than polar.

In this exercise, you will draw the Lewis structures of a set of molecules. Then you will use a molecular model kit to build the molecules. Finally, you will describe each molecule's molecular geometry and state whether each molecule is polar or nonpolar.

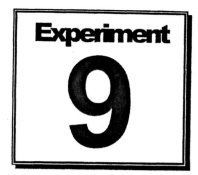

Solubility of Polar and Nonpolar Compounds

Like dissolves like.

One of the most important physical properties of a chemical compound is its solubility. When one compound is soluble in another, they form a solution. A **solution** is a homogeneous mixture that generally contains a large amount of one component (**solvent**) into which a small amount of another component (**solute**) is dissolved.

Miscibility is a special case of solubility. When two liquids are **miscible** they will mix in all proportions, or are said to be infinitely soluble.

You have probably observed that some pairs of liquids, such as oil and water, will not mix, whereas others, such as alcohol and water, will mix readily. Two compounds that will dissolve in one another are said to be **soluble** and form a solution as described above. Those that will not dissolve are referred to as **insoluble**. The common axiom used to describe this is "like dissolved like."

Solubility of liquids (or lack thereof) can be explained by a simplified discussion of polarity. In chemistry, we generally divide compounds into two large groups: polar and nonpolar. In biological systems, we talk of chemicals as being water (polar) or fat (nonpolar) soluble. Polarity is based on the distribution of electrical charge on the molecule of the compound. Molecules that have specific areas of positive and negative charge are polar. Water and alcohol are examples of polar molecules. In the molecules of a nonpolar material, the charge is evenly distributed; therefore, there are no positive and negative regions. Because the magnitude of charge distribution can vary, a range of polarities can exist. It is important to remember that the total charge on any *molecule* equals zero. In polar molecules, however, the positive and negative charges are each relegated to distinct regions. The extreme of distributed charges is found in ionic compounds, where you have distinctly separate ions with positive and negative charges.

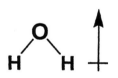

A water molecule is polar.

Although both sugar and table salt readily dissolve in water, there is a significant difference in the two solutions. The salt solution is able to conduct electricity, whereas the sugar solution does not. This ability to conduct electricity when dissolved in water is characteristic of soluble ionic compounds. When ionic compounds dissolve in water, they form a solution that contains the separate ions. These solutions conduct electricity, and the dissolved compounds are termed **electrolytes**. Some covalent compounds form ions in solution and are also electrolytes. You can test for the presence of electrolytes by using a conductivity tester.

Electrolytes are substances that produce ions in solution.

Electrolytes are significant in biological systems because they are important in maintaining cell performance. Particular ions are required for specific functions in the body. Sodium and potassium ions are the major cations present in extracellular and intracellular fluid, respectively. There are many cases where you need to supply electrolytes to your systems. When there is a loss of fluids and/or electrolytes, a solution may be administered to replenish them. Drinks such as Powerade™ and Gatorade™ provide electrolytes and carbohydrates during exercise. Babies and young children can become dehydrated and lose electrolytes when ill with diarrhea and vomiting. Doctors often prescribe solutions, such as Pedialyte™, to supply fluids and electrolytes.

Solubility is also affected by temperature. Most solids have an increased solubility at higher temperatures. Temperature has a large effect on the solubility of gases in water. As the temperature increases, gas is less soluble in water. You notice this effect in everyday life when you leave a soft drink out at room temperature. As the temperature of the drink increases, the carbon dioxide gas (the carbonation) escapes and your pop loses its "fizz."

MATERIALS

Equipment
- small test tubes
- test-tube rack
- spatula
- wood splints
- 10-mL graduated cylinder
- dropper
- beakers
- hot plate or heat source
- conductivity tester
- thermometer
- marking pencil
- 100-mL graduate cylinder
- 100-mL beakers (4)

Chemicals
- sucrose, $C_{12}H_{22}O_{12}$
- iodine, I_2
- ammonium chloride, NH_4Cl
- calcium chloride, $CaCl_2$
- sodium chloride, $NaCl$
- naphthalene flakes, $C_{10}H_8$
- potassium permanganate, $KMnO_4$
- ethanol, C_2H_5OH
- vegetable oil
- hexane, C_6H_{14}
- bromthymol blue
- Perrier or other carbonated bottled water

CAUTION!
Iodine can irritate the skin and eyes; handle it with particular caution.

WASTE DISPOSAL
Both hexane and ethanol are flammable. These liquids cannot be poured down the drain. Place any excess or waste liquids and mixtures into the appropriate waste containers.

PROCEDURE

Part A: Solubility in polar and nonpolar solvents

Don't forget to put on your goggles!

1. Label nine test tubes 1 through 9.

2. Using a graduated cylinder, measure and pour 3 mL of water (distilled, if available) into a test tube. Mark the height of this volume on test tube #1 with a marking pencil. Place a mark at the same height on the nine other labeled test tubes. Add 3 mL of water to each test tube by filling up to the mark.

3. Test the water solubility of each of the ten solutes by adding a different solute to each test tube. The ten solutes are: sucrose, iodine, ammonium chloride, calcium chloride, sodium chloride, naphthalene, potassium permanganate, ethanol, hexane, and vegetable oil. For liquid solutes, add approximately 1 mL. Transfer a small amount (about the size of small pea) of each solid solute, using a metal spatula or wood splint. Because iodine reacts with some metals, you must use a wood splint for iodine; discard the splint after use.

When some compounds are dissolved in water, the resultant solution can be quite warm or cold. This is the basis for some hot and cold packs.

4. Gently mix each test tube's contents by firmly tapping on the side of the tube.

5. Make judgments about the solubility of each solute: soluble, slightly soluble, insoluble. Also, note any changes in color, temperature, etc. For liquids note whether the solute was less dense (is on top) or more dense than water.

6. Test the conductivity of pure water. Insert the probes of the conductivity tester into the water and note any illumination of the light emitting diode (LED). Distilled water should not conduct electricity. If you used tap water, note the brightness of the LED. You will be able to make comparative measurements of conductivity. Anything "brighter" than pure water means that there are some electrolytes present.

7. Test the conductivity of your solutions. If the probes do not reach down to the solution, transfer your solution to a small beaker for the test. Remember to rinse the probes after each test.

8. Discard all test tube contents into labeled waste containers.

9. Wash and thoroughly dry the test tubes.

10. Repeat Steps 2 through 5, using the nonpolar solvent hexane instead of water.

Part B: Temperature effect on the solubility of gases

Many bottled waters and soft drinks are carbonated because they contain dissolved carbon dioxide gas. Carbon dioxide forms an acidic solution when dissolved in water, as a result of the formation of carbonic acid.

$$H_2O + CO_2 \rightleftharpoons \underset{\text{Carbonic acid}}{H_2CO_3} \rightleftharpoons H^+ + HCO_3^-$$

If the solubility of carbon dioxide gas changes with increasing temperature, you should see a change in the acidity of a carbonated solution upon boiling. For this part of the experiment, you will use an indicator (bromthymol blue) that is yellow at low pH (acid solution) and blue at high pH (neutral or basic solutions).

1. Open a bottle of Perrier water and dispense 30 mL into each of two beakers.

2. Add three drops of bromthymol blue indicator to each beaker. Note the color of the solution.

3. Heat one of the beakers on a hot plate for 5–10 minutes (or to boiling). Allow the solution to cool. Note the color of the solution after heating.

Part C: Temperature effect on solubility of solids

In contrast to gases, the solubility of most solids increases with temperature. You know from experience that sugar dissolves more readily in hot tea than in iced tea. We will test this in the following part of the experiment.

1. Fill two 100-mL beakers with water. Heat one beaker to 70°–80°C.

2. Add several (4–5) crystals of potassium permanganate ($KMnO_4$) to each beaker.

3. Describe the intensity of the color in each beaker over time: immediately after addition, 1 minute, 5 minutes, up to 15 minutes.

DATA

Classify the solutes as soluble, slightly soluble or insoluble.

Solute	Solubility in water (H_2O)	Conductivity	Solubility in hexane (C_6H_{14})
sucrose, $C_{12}H_{22}O_{11}$			
iodine, I_2			
ammonium chloride, NH_4Cl			
calcium chloride, $CaCl_2$			
sodium chloride, NaCl			
naphthalene, $C_{10}H_8$			
potassium permanganate, $KMnO_4$			
ethanol, C_2H_5OH			
hexane, C_6H_{14}			
vegetable oil			

OBSERVATIONS

Solubility in polar and nonpolar solvents

Make notes about any color, temperature, or other changes you observed when making the solutions.

Temperature effect on the solubility of gases

Describe any color changes observed in the indicator in the Perrier water.

Temperature effect on the solubility of solids

Describe your observations on the solubility of potassium permanganate.

Time	Room temperature	Heated water
0 minutes		
5 minutes		
15 minutes		

QUESTIONS

1. Classify each of the solutes as covalent or ionic (yes/no).

Solute	Covalent	Ionic
sucrose, $C_{12}H_{22}O_{11}$		
iodine, I_2		
ammonium chloride, NH_4Cl		
calcium chloride, $CaCl_2$		
sodium chloride, $NaCl$		
naphthalene, $C_{10}H_8$		
potassium permanganate, $KMnO_4$		
ethanol, C_2H_5OH		

NAME _____

Solute	Covalent	Ionic
hexane, C_6H_{14}		
vegetable oil		

2. Which solutes would you classify as nonpolar?

3. What property do the covalent solutes have in common?

4. Acetic acid is a covalent compound but, when dissolved in water, it is a weak electrolyte. How would you explain this phenomenon?

5. Did you see any change in the color of the indicator in the Perrier water after heating? Explain this result.

Experiment 10

Paper Chromatography

Chromatography is a versatile separation method.

Paper chromatography is a technique often used by chemists to separate components of a mixture. In paper chromatography, a small drop of the mixture is applied near one end of a strip of paper or **stationary phase**. The end of the paper is then dipped into a developing solvent, often called the **mobile phase**, which flows up the paper by capillary action. As the developing solvent flows up the paper, it can carry along the components of the mixture. The rate at which a particular component moves depends on whether it tends to be dissolved in the developing solvent or whether it prefers to remain absorbed (or stuck) on the surface of the paper. A component that moves rapidly is spending more time dissolved in the developing solvent, the mobile phase; whereas a component that moves slowly is spending more time adsorbed to the paper, the stationary phase.

The developing solvent must be chosen carefully to achieve optimal separation of the components in the mixture. Selection of the solvent is the first step in a chromatographic technique. In the first part of this experiment, you will first identify the solvent that gives the best separation of the inks used in the transparency pens. You will then use this solvent to compare the inks used in several different pens.

In the second part of the experiment, you will chromatograph a mixture of $Cu(NO_3)_2$, $Ni(NO_3)_2$, and $Fe(NO_3)_3$ along with an unknown mixture of two or three of these salts. Because the cations Cu^{2+}, Ni^{2+}, and Fe^{3+} have different properties, they will have different interactions with the mobile and the stationary phases. Therefore, various ions will travel up the paper at different speeds. The identity of components in an unknown mixture can be deduced by comparing the chromatogram of the unknown with the chromatogram for a known.

Not all the cations are visible on the paper, so they will be discerned in the following manner:

- Fe^{3+}, iron(III) ion, in water imparts a rust color and thus will produce a faint rust-colored band on the paper.

- Cu^{2+}, copper(II) ion, is blue in water, but the color is too faint to produce a blue band on the paper. However, Cu^{2+} reacts with NH_3 (from ammonium hydroxide) to form a complex ion, $[Cu(NH_3)_4]^{2+}$, which is deep blue and is readily observed.

- Ni^{2+}, nickel(II) ion, which is green in water, will react with an organic reagent, dimethylglyoxime (DMG), to produce a strawberry red color.

The results of a chromatographic separation are often expressed in terms of R_f values. An R_f is the *relative* distance that a sample component has moved. You measure the distance from the point where you applied each sample (the origin) to the middle of each component spot and the distance from the origin to the location of the solvent front.

A diagram of a typical chromatogram is shown below.

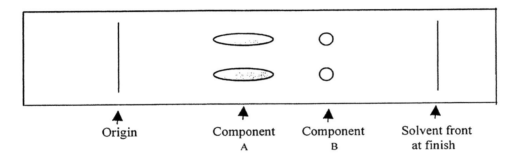

For example, if component A traveled 4 cm and the developing solvent traveled 10 cm, then you can calculate the R_f as follows:

$$R_f(A) = \frac{\text{distance A traveled}}{\text{distance solvent traveled}} = \frac{4 \text{ cm}}{10 \text{ cm}} = 0.4$$

R_f values determined under identical conditions are reasonably constant. In other words, your calculated R_f values can be compared with another person's data. Note that the R_f value will change if you use a different solvent.

MATERIALS

Equipment

- test tubes
- test-tube rack
- 800-mL beaker
- 10-mL graduated cylinder
- grease pencil
- chromatography paper
- transparency pens
- plastic wrap
- stapler
- corks
- glass rod
- capillary tubes
- paper clips

Chemicals

- ethanol
- acetone
- 0.5 M copper sulfate, $Cu(NO_3)_2$
- 0.5 M nickel nitrate, $Ni(NO_3)_2$
- 0.5 M iron (III) nitrate, $Fe(NO_3)_3$
- 1% dimethylglyoxime (DMG) in ethanol
- conc. ammonia
- 9:1 acetone:6 M HCl
- cation unknowns

SAFETY/DISPOSAL

Both acetone and ethanol are flammable. These solvents cannot be rinsed down the drain. Place any excess or waste developing solvents in the marked waste containers.

PROCEDURE

Part A: Chromatography of ink

1. Label four test tubes 1 through 4 with a grease pencil. Get a #10 cork stopper for each test tube.

2. In each tube, pour the solvent listed below to a depth of approximately 1 cm. Stopper each tube and set each in a test-tube rack.

Test tube #	Solvent
1	acetone
2	distilled water
3	ethanol : water, 1:1 mixture*
4	ethanol

    *The ethanol : water mixture is prepared by thoroughly mixing 3 mL of each solvent.

3. Draw a pencil line (NOT INK) about 2 cm from the bottom of four pieces of narrow chromatography paper. Using the black transparency pen, go over the

pencil line. Let dry. Label the top of the papers 1 through 4 with a pencil. Fold the pieces of paper in half lengthwise.

4. Place chromatogram #1 in tube #1, chromatogram #2 in tube #2, etc. Be sure the ink line is not immersed in the solvent.

5. Stopper the tubes and let the solvent move up the paper. After the solvent has moved about 10 cm up the paper, remove the paper and mark the solvent front (the highest point traveled by the solvent) with a regular lead pencil.

6. Let the chromatograms dry, record your observations, and then decide which solvent gives the best separation.

7. Pour 30 mL of the selected solvent into an 800-mL beaker and cover the beaker with plastic wrap. Use a wet paper towel to moisten the outside of the beaker and seal the plastic wrap to the moistened area.

8. Draw a pencil line about 1 cm from the bottom of a sheet of chromatography paper (approximately 20 cm × 10 cm) and draw a pencil line about 1 cm from the bottom of this paper. At several places on the pencil line, make a short pen line with the transparency pens provided.

9. Let the ink dry, then bring the two short ends of the chromatogram together to form a cylinder. Staple the ends, but do not overlap the ends.

10. Place the chromatogram in the beaker and replace the plastic wrap on top, sealing it to the outside of the beaker.

11. When the solvent front is within 1 to 2 cm of the top of the paper, remove the paper from the beaker and mark the solvent front with a lead pencil. Set the paper aside to dry.

12. Once the paper is dry, observe the colors separated in each pen and record your observations.

13. Attach your chromatogram to the report.

Time Saving Advice!

While the chromatogram is running, you can begin Part B of the experiment.

Part B: Chromatography of Cations

1. Using a pencil, mark an "origin" line 2 cm from the short end of a 6 cm × 10 cm piece of chromatography paper. Using a capillary tube, spot the paper on the starting line 1 cm from the long edge with the known mixture of Cu^{2+}, Fe^{3+}, and Ni^{2+}. Allow the spot to dry and repeat the application at least 2 more times. It is very important to get a concentrated spot on your paper or you might not be able to detect the metal ion after developing the chromatogram. At distances of at least 1 cm away from the spot for the known mixture (but still on the origin line) and

from each other, spot two of the unknown mixtures. Allow the spots to dry. You should have three spots: the known mixture and two unknowns.

2. To a clean 800-mL beaker, add developing solvent to a depth of approximately 6 mm. The developing solvent consists of a 9:1 acetone : 6 M HCl mixture and may be somewhat yellow in color. Cover the beaker with plastic wrap. Use a wet paper towel to moisten the outside of the beaker and seal the plastic wrap to the moistened area.

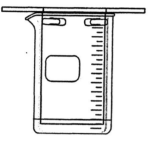

3. Attach the chromatogram to a glass rod by folding the unspotted end (not the end with the origin line) over the glass rod and secure it with a paper clip. See the diagram in the margin. Now carefully remove the plastic wrap from the 800-mL beaker and place the glass rod across the top of the beaker so that the chromatography paper is barely dipping into the developing solvent. Do not immerse the starting line in the solvent. Carefully recover the beaker with a piece of plastic wrap.

4. When the solvent has nearly reached the union of the folded part of the paper, carefully remove the plastic wrap and lift out the chromatogram. Immediately mark how far the solvent has moved (the solvent front), with a pencil. Allow the chromatogram to air dry.

5. Once the chromatogram has dried, inspect the chromatogram for the rust color of Fe^{3+}. The copper spot may also appear yellow at this point, so don't mistake this for the Fe^{3+} spot. If you are unsure, circle both spots. You will determine which is copper in the next step.

CAUTION!
Ammonium hydroxide vapors are irritating. Do not inhale. Only use in the hood.

6. In the hood, you will find an Erlenmeyer flask containing concentrated ammonium hydroxide. Expose your chromatogram to the ammonium hydroxide vapors by moving the chromatogram back and forth above the mouth of the flask for 20–30 seconds. Do not allow the chromatogram to dip into the ammonium hydroxide solution, only expose the chromatogram to the vapor. A deep blue color will develop if copper ion is present. If you don't get any blue color (even for the standard mixture), just move on to the next step. An added benefit of this step is that it may intensify the rust color of the Fe(III)

7. Paint the dimethylglyoxime (DMG) solution onto the chromatogram by using a glass rod that has been dipped in the DMG solution. A strawberry red color will develop if nickel ion is present. At this point, you may need to repeat Step #6 (expose the chromatogram to ammonia vapors) in order to see the copper spots.

8. Circle each spot, using a pencil. Measure the distance traveled from the origin by each cation in the known and unknown mixtures. Draw a diagram of the chromatogram.

NAME _____

DATA

Part A: Chromatography of ink

1. Solvent with best separation is _____. What basis did you use to decide on this solvent?

2. Attach the chromatogram for the different colored pens. State and discuss your results.

Part B: Chromatography of cations

1. Draw a diagram of chromatogram of the cations.

2. Record the distance traveled by solvent, each cation and the unknown.

   Distance traveled by solvent   _____

   Distance traveled by iron(III)   _____

   Distance traveled by copper(II)   _____

   Distrance traveled by nickel(II)   _____

RESULTS

Calculate the R_f values of known cations. Show at least one sample calculation below.

$$R_f = \frac{\text{distance cation traveled}}{\text{distance solvent traveled}}$$

Sample	R_f value	Identity of cation
Known mixture iron(III) nickel(II) copper(II)		
Unknown 1		
Unknown 2		

State the basis for your unknown identification.

NAME _____

QUESTIONS

1. Why is it essential that the application spot in paper chromatography be above the surface of the solvent?

2. How would your analysis be affected if the paper chromatogram was not removed from the developing solution when the solvent reached the top of the paper?

3. According to the data from Part B, which cation had the strongest affinity for the stationary phase? Which cation had the strongest affinity for the mobile phase? Explain your answer.

Experiment 11

Stoichiometry and the Chemical Equation

The reaction of hydrogen peroxide and bleach.

> If you mix bleach and ammonia together, the toxic gas chlorine can form.

You would be surprised at the number of chemical reactions you can do with common household products. In fact, you need to be very careful when working with some cleaning agents because they can react to form toxic gases. In this experiment, you will be mixing two common household products, but the product in this case is a nontoxic gas, oxygen. When hydrogen peroxide (H_2O_2) is mixed with bleach, bubbles of oxygen gas are formed. The active ingredient in bleach is sodium hypochlorite, $NaOCl$, which is present as a 5.25% solution in water. In this experiment, you will measure how much oxygen is produced when known amounts of hydrogen peroxide solution are mixed with known amounts of bleach. Based on the amount of reactants used (bleach and hydrogen peroxide) and the amount of product formed (oxygen), you will determine the stoichiometry of the chemical reaction.

> The **limiting reagent** restricts the amount of product formed.

Stoichiometry is the study of the quantitative relationship between substances undergoing chemical reactions. You may have encountered similar sorts of problems outside the lab— for instance, figuring out how many cookies (products) can be made from a given amount of ingredients (reactants) based on a specific recipe (equation). If a recipe requires that 2 cups of flour be mixed with 4 eggs to make 24 cookies, how many cookies can be made from 1 cup of flour and 2 eggs? Like the ratio of the ingredients in cooking, the ratios of the reactants in a chemical reaction are fixed. So how many cookies can you make if you have 1 cup of flour and a dozen eggs? It is obvious that the number of eggs is in excess, or you might say that the number of cookies you can make is limited by the amount of flour. In chemistry, the reagent that is completely used up in a reaction is called the **limiting reagent**.

The reaction between bleach and hydrogen peroxide acts as if it were following a recipe. How will the volume of oxygen produced from 4 mL of hydrogen peroxide vary with the amount of bleach added to it? Make a prediction. Did your reasoning include the fact that the amount of oxygen gas produced is dependent on both the amounts of hydrogen peroxide and bleach? These two compounds react according to specific mole ratios that you will be determining from the results of this experiment.

MATERIALS

Equipment

- large beaker for waste collection
- small beakers (2)
- 100-mL graduated cylinder
- 10-mL graduated cylinder (2)
- tweezers
- small vial
- rubber tubing
- rubber stopper
- 250-mL side-arm Erlenmeyer flask
- ring stand with clamp
- water trough

Chemicals

- 3% aqueous hydrogen peroxide, H_2O_2
- Bleach, 5.25% aqueous NaOCl

PROCEDURE

You will run two sets of reactions for this experiment. In Reaction Set A, the volume of hydrogen peroxide will remain constant (4 mL) and the volume of bleach will vary. In Reaction Set B, the volume of bleach will remain constant (4 mL) and the volume of hydrogen peroxide will vary. The volumes of reactants to be used for each set of reactions are given below. This may seem like a lot of runs but each run takes just minutes to perform.

Set A	Volume of bleach	Volume of H_2O_2
Run 1	4 mL	1.5 mL
Run 2	4 mL	2.5 mL
Run 3	4 mL	3.5 mL
Run 4	4 mL	4.5 mL
Run 5	4 mL	5.5 mL
Run 6	4 mL	6.5 mL
Run 7	4 mL	7.5 mL

Set B	Volume of Bleach	Volume of H_2O_2
Run 1	1.5 mL	4 mL
Run 2	2.5 mL	4 mL
Run 3	3.5 mL	4 mL
Run 4	4.5 mL	4 mL
Run 5	5.5 mL	4 mL
Run 6	6.5 mL	4 mL
Run 7	7.5 mL	4 mL

1. Fill a water trough with tap water. Completely immerse a 100-mL graduated cylinder (if necessary, remove the plastic bottom from the cylinder) in the water trough, filling it with water. Turn the cylinder upside down, keeping the mouth below the surface of the water in the trough. Clamp the cylinder onto a ring stand positioning it high enough to allow you to slip a piece of rubber tubing into the mouth of the cylinder. The up-ended graduated cylinder should be full of water (NO AIR BUBBLES) and securely fastened onto the ring stand. This is the oxygen-measuring vessel; gas formed by the reaction will bubble into the up-ended cylinder, where it will displace some of the water. Then you will be able to read the volume displaced directly from the graduations on the cylinder. The overall setup for the gas-collecting apparatus is shown below.

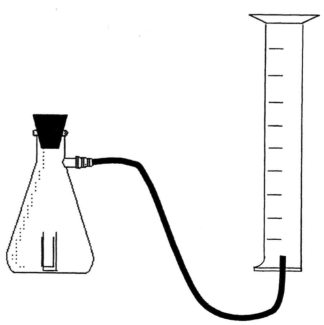

2. Label two clean, dry small beakers, one for bleach and one for hydrogen peroxide. Obtain approximately 45 mL of bleach and 45 mL of hydrogen peroxide.

3. Label the two 10-mL graduated cylinders, one for bleach and one for hydrogen peroxide. With the aid of a disposable pipette, transfer the designated amount of bleach (4 mL for Set A, reaction 1) into the 10-mL graduated cylinder. Because this experiment is to be quantitative, it is necessary to measure the quantities for each run exactly. Pour the bleach into the side-arm Erlenmeyer flask.

4. Measure out the designated amount of hydrogen peroxide (1.5 mL for Set A, reaction 1) into a 10-mL graduated cylinder. Pour the hydrogen peroxide into the small vial, then use tweezers to lower the vial into the sidearm flask, taking care not to knock over the vial.

5. Again being careful not to knock over the vial, stopper the flask with the rubber stopper. Push it in firmly to form a good seal. Place the loose end of the tubing into the mouth of the up-ended graduated cylinder (refer to diagram above). You may have to hold the tubing to prevent it from flopping around. Don't worry if a bubble or two escapes from the tubing into the cylinder.

6. Once the setup is complete, jostle the reaction flask until the vial tips over and spills the hydrogen peroxide into the bleach. Swirl the flask to ensure complete mixing. Oxygen should begin forming immediately. Some reactions may finish quite rapidly; others may take several minutes. Wait until the mixture in the flask stops fizzing and oxygen stops bubbling into the graduated cylinder. Record the amount of gas that was produced by noting the level of the liquid-gas interface in the graduated cylinder. Remember that the cylinder is upside down when you read the markings. The volume of oxygen produced should be recorded in the data table in the Data section.

7. Empty the flask by pouring the reaction solution into the large beaker that will serve as your waste container. Rinse the flask and vial several times with water. The flask and the vial do not have to be dry for the subsequent runs.

8. Repeat all of the reactions noted above for Reaction Set A and Reaction Set B. Record the volume of oxygen produced in each run.

Charles's Law

Charles's law describes the relationship between the volume and the temperature of a gas.

In 1787, Jacques Charles, a French scientist, investigated the effect of changing the temperature of a given amount of air while holding the pressure constant. He found that when room temperature air was heated to 100°C, it volume increased; whereas when the air was cooled to 0°C, its volume decreased. This relationship between volume and temperature is direct and linear.

When the value of one factor increases with the increase of another, the relationship is called direct.

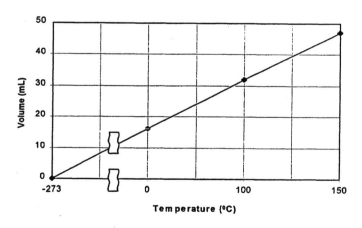

Volume of an enclosed gas as a function of temperature at constant pressure

Because the volume of gas decreases in a linear fashion from 100°C to 0°C, we expect it to continue to decrease in a linear fashion below 0°C. This expectation leads us to the conclusion that, because we cannot have a negative volume, we cannot obtain a lower temperature than that for which the volume is zero. This temperature is called the **absolute zero** of temperature and can be found (theoretically) by extending the line (extrapolating) for the volume-temperature data to a volume of zero. That temperature is has been determined to be −273.15°C. This value is defined as the zero point on the Kelvin (K) temperature scale.

Useful conversion
0 K = −273.15°C

The relationship of these factors is stated in Charles's law as follows: The volume of a fixed amount of gas maintained at constant pressure is directly proportional to its absolute temperature. This relationship is expressed symbolically as $V \propto T$ or mathematically as $V = cT$, where c is a proportionality constant dependent on the pressure and the amount of gas. In these equations, the temperature must be expressed in kelvins.

In this experiment, you will examine the temperature-volume relationship discovered by Jacques Charles and use your data to determine the Celsius equivalent of an absolute zero temperature. You will also determine the value of c.

MATERIALS

Equipment

- ✓ Bunsen burner
- ✓ ring stand
- ✓ clamps
- ✓ 600-mL beakers (2)
- ✓ thermometer
- ✓ syringe
- ✓ syringe cap

Chemicals

- ✓ ice water

PROCEDURE

1. Pull out the plunger of the syringe so that the syringe contains about 20 cc of air. Use the syringe cap to plug the hole in the tip of the syringe. Record the exact volume and the room temperature.

2. Add approximately 400 mL of water to each beaker. Insert the syringe into the water to test the water depth. You want the water to completely cover the barrel of the syringe to at least 30 cc.

3. Allow the water in one beaker to come to room temperature while you set up and heat the water in second beaker to boiling.

4. Clamp the syringe, thermometer, and water bath to the ring stand as shown in the picture below. Be sure there is sufficient water in the beaker to completely submerge the air contained in the syringe. Heat the water to a gentle boil by using

a Bunsen burner. Allow about 2 minutes for the syringe and air to attain the same temperature as that of the water and then record the temperature of the water and volume of the air in the syringe.

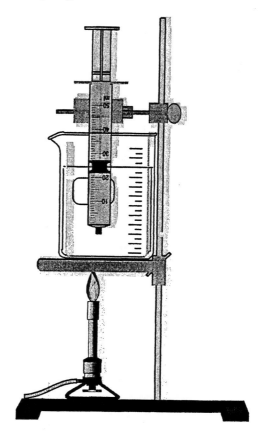

5. Turn off the Bunsen burner and carefully move the thermometer and the syringe to the other 600-mL beaker containing water at room temperature. Allow about 5 minutes for the syringe of air to attain the same temperature as the water. Record the temperature of the water bath and the volume of the air in the syringe.

6. Add a few small pieces of ice to the room-temperature beaker to lower the temperature of the water to approximately 10°–15°C. Allow time for the temperature of the water to stabilize and the air in the syringe to attain the same temperature. Record the temperature of the water bath and the volume of the air in the syringe.

7. Pour off half of the 10°–15°C water from the beaker and add more ice. Allow about 5 minutes for the syringe of air to attain the same temperature as the ice/water bath and then record the temperature of the water bath and the volume of the air in the syringe.

Electrophoresis of Dye Molecules

Electrophoresis is a common technique used in biotechnology.

Electrophoresis, the transport of particles by an electrical field, can be used to separate molecules on the basis of charge. Electrophoresis is usually carried out in aqueous solutions containing buffer ions and salts in addition to the molecules of interest. The charged particles move through a liquid that conducts an electric current. Molecules with a net negative charge will migrate toward the positive electrode; molecules with a net positive charge will migrate toward the negative electrode.

Different molecules will move through the electric field at different rates. Charge is only one variable to consider when looking at the movement of a molecule in an electric field. The velocity of migration (v) of a molecule in an electric field depends on the electric field strength *(E)*, the net charge on the molecules *(z)*, and the frictional coefficient *(f)*:

$$v = \frac{Ez}{f} \quad \text{or} \quad fv = Ez$$

The electric force (*Ez*) driving the charged molecule toward the oppositely charged electrode is opposed by the viscous drag (*fv*) arising from the friction between the moving molecule and the medium. The frictional coefficient depends on both the mass and shape of the migrating molecules and the viscosity of the medium through which the molecules are moving. Larger molecules have a larger frictional coefficient. Even though the equation may seem complex, it just states that large molecules will move slower than small ones, and more highly charged molecules will move faster than less highly charge ones. Therefore, when looking at the migration of molecules in electrophoresis, you must consider net charge, molecular weight, and molecular shape.

Generally, electrophoresis is carried out not in free solution but in a porous gellike support matrix such as polyacrylamide or the polysaccharide agarose. These gels retard the movement of molecules. Larger molecules are retarded in their migration through the gel pores, whereas the movement of smaller molecules is relatively unhindered.

The apparatus required for electrophoresis is actually quite simple. The two main components of any electrophoresis systems are a power supply and the gel box. The power supply is the source of the direct electric current [it converts the alternating current (AC) supplied by the wall socket into a direct electric current (DC)]. The gel box holds the support matrix and a conducting solution. The gel box also contains two electrodes. The electrode where the electrons enter the gel box from the power supply is called the cathode (**black**) and is **negative**. The electrode where the electrons leave the gel box and return to the power supply is called the anode (**red**) and is **positive**.

When the current is flowing, there are chemical reactions occurring at the cathode and at the anode. These reactions form hydroxide ion (OH^-) at the cathode and hydronium ion (H_3O^+) at the anode. Because of these products, it is important to use buffer solutions during electrophoresis. The accumulation of hydroxide ions at the cathode and hydronium ions at the anode would lead to significant pH changes if the solution were not buffered.

> A **buffered solution** is a solution that resists changes in pH when hydroxide ion or hydronium ion is added.

Electrophoresis is one of the most important techniques used in biochemistry. It can be used to assess the molecular weight of a protein or genetic material, such as DNA. It is also the approach of choice for assessing the number of proteins or DNA fragments in a mixture. In this experiment, you will use electrophoresis to separate dye molecules and to investigate the effects of size and charge on electrophoretic mobility. Dyes may be classified by their chemical structure, with the dye existing as a neutral molecule or as an ion in the cationic (+ charged) or the anionic (− charged) form.

MATERIALS

Equipment
- DC power supply
- gel box with leads, comb, and dams
- pipettors, 2–20 µL (2)
- 10-mL graduated cylinder
- 100-mL graduated cylinder
- watchglass
- 100-mL beaker
- 250-mL Erlenmeyer flask
- stirring rod
- microtubes

Chemicals
- 50× TAE buffer (TAE = 1 M Tris base, 1 M acetic acid, 0.05 M EDTA)
- various dyes
- agarose powder

PROCEDURE

Part A: Preparations of 1% agarose gel

1. Prepare 200 mL 1× TAE buffer by taking 4 mL of 50× buffer and diluting to 200 mL with distilled H_2O.

2. Weigh out 0.5 g of agarose and place it in a beaker. Add 50 mL of 1× TAE buffer. Cover your beaker with a watchglass and heat this mixture in a microwave for 2 minutes until all the agarose particles are dissolved. Use the thermal gloves to remove the beaker from the microwave.

> **CAUTION!**
> Beaker and agarose will be HOT. Use thermal gloves.

3. Set up the gel box to receive agarose by placing dams at each end of the gel deck.

4. Pour sufficient slightly cooled agarose solution into the gel tray to just fill the tray up to the edge. Make sure that there are no bubbles in the molten agarose. If there are any bubbles in the gel, gently shake the gel bed. Insert the comb into the middle slot. Bubbles must be removed and comb inserted before the gel solidifies.

5. Let the gel harden for 10 minutes. Add 125 mL of 1× TAE buffer (prepared in step 1), dividing the solution between each side of the gel box. Remove the comb and dams gently. The gel should be fully submerged in the buffer. Give the gel box a gentle shake to make sure the wells contain buffer. Add more buffer if needed.

Part B: Running the gel

1. Number a set of microtubes 1 through 10. Make a list of 10 dyes and assign each a number.

2. Pipet 17 µL of each dye into the appropriately numbered microtube. This quantity will allow you to load a 15-µL sample.

3. Before loading your sample, move your gel box near the power supply you are going to use. You will not want to move your gel box once the dyes are loaded.

4. Tap the microtubes on the bench top to collect any droplets of sample into the bottom of the microtube. Load 15 µL of sample into each well. To load the sample, dip the loaded pipet tip through the surface of the buffer, centering it over the first well. The tip should just barely enter the well. Slowly and steadily, depress the plunger. You should see the sample sink neatly into the well. Do not inject air into the well, because this will cause the dye to come swirling out of the well. Record which sample is loaded into each well. Be careful when loading the gel to place the pipet tip into the well without puncturing the bottom of the well.

5. When all samples are loaded, close the lid on the gel box and connect it to the power supply. If you are sharing a power supply, make sure the power supply is turned off when connecting your gel box. Set the power supply to the setting indicated by the teaching assistant.

6. Electrophorese for at least 15 minutes or longer for better separation.

7. After 15 or more minutes, turn off the power and unplug the electrodes. Gently remove the gel from the tank and place it on the lab bench.

DATA ANALYSIS

1. Draw a diagram of the gel below. Label the wells and each dye spot and the (+) and (−) ends.

2. Classify the dyes you used as either cationic or anionic.

   Cationic dyes   Anionic dyes

3. Look up the structure of two of the dyes you used and draw them below. Can you rationalize the migration of these dyes on the basis of their structure?

Reactions of Common Metals

A look at the various chemical reaction types

Redox Reactions

One of the most important characteristics of a metal is its activity (its reactivity), or its ability to lose electrons and become an ion. Metals range widely in their activity, from vigorously reactive cesium, potassium, and sodium to quite inactive platinum, gold, and silver. In this experiment, you will rank some metals according to their activities, from most active to least active. This ranking of metal according to reactivity is called an **activity series**. In performing this experiment, you will observe whether an oxidation-reduction (redox) reaction occurs between a metal and the cation of another metal. **Oxidation reactions** involve the loss of elections; **reduction reactions** involve the gain of elections. These reactions occur as a set: to lose electrons (oxidation), there must be another reactant that accepts those electrons (reduction).

The set of reactions involving oxidation and reduction are called redox reactions.

A redox reaction occurs when copper metal is place in a solution of silver nitrate. During this reaction, crystals of silver metal appear and the solution turns blue because of the copper ions formed in the reaction. This reaction can be written:

$$Cu\,(s) + 2\,AgNO_3\,(aq) \longrightarrow Cu(NO_3)_2\,(aq) + 2\,Ag\,(s)$$

In this example, copper is said to be more active than silver because it donates its electrons to silver ions. A more active metal will reduce (donate electrons) a less active metal. To show exactly where electrons are lost and gained, the equation can be divided into two half-reactions:

$$Cu^0 \longrightarrow Cu^{2+} + 2\text{ electrons (copper loses electrons; it is oxidized)}$$

$$Ag^+ + 1\text{ electron} \longrightarrow Ag^0\text{ (silver ion gains an electron; it is reduced)}$$

What would you expect if instead of copper metal placed in silver nitrate solution, silver metal were placed in copper nitrate solution? No reaction occurs, because silver is less active than copper. With this in mind, we can rank the activity of copper and silver as

    More active   copper
    Less active   silver

Acids may also participate in redox reactions with active metals to produce hydrogen gas and a salt. For example, aluminum reacts with hydrochloric acid:

$$2\ Al^0\ (s) + 6\ HCl\ (aq) \longrightarrow 2\ AlCl_3\ (aq) + 3\ H_2\ (g)$$

Silver, on the other hand, does not react with hydrochloric acid:

$$Ag^0\ (s) + HCl\ (aq) \longrightarrow \text{no reaction}$$

Metals that are more active than hydrogen (like aluminum) will displace hydrogen from acids, and those less active (like silver) are unable to do so. With this in mind, we can rank the activity of aluminum, hydrogen, and silver as

    Most active   aluminum
                             hydrogen
    Least active   silver

In the first part of this experiment, you will use the information from displacement reactions and the reactions with acid to rank the activity of zinc, magnesium, copper, and hydrogen.

Precipitation Reactions

Another common reaction of metals and their ions is a precipitation reaction. Ionic solids are made up of individual ions held together by electrostatic forces. When such solids are dissolved in water, they dissociate and the ions separate and become essentially independent of one another. Some ionic solids have a limited ability to separate in solution. When the ions of such solids are mixed in solution, they "fall" out of solution as a precipitate.

An ion in solution may be precipitated by the addition of a compound that reacts with the ion to form an insoluble precipitate.

In this experiment, you will determine the behavior of a number of different kinds of metal ions and then use this information to analyze an unknown substance. You will study the metal ions Ba^{2+}, Cu^{2+}, Pb^{2+}, and Ni^{2+}. To analyze these metal ions, you will use the anions carbonate (CO_3^{2-}), hydroxide (OH^-), iodide (I^-), and sulfate (SO_4^{2-}). Each of these anions will be mixed with each of these cations, and you will observe whether a precipitation reaction takes place.

MATERIALS

Equipment

- ✓ small test tubes
- ✓ test tube rack
- ✓ 24-well plate

Chemicals

- ✓ zinc, magnesium, and copper metal
- ✓ 0.1 M Na_2CO_3
- ✓ 0.5 M $CuSO_4$
- ✓ 0.5 M $ZnSO_4$
- ✓ 0.5 M $MgSO_4$
- ✓ 6 M NaOH
- ✓ 0.1 M NaI
- ✓ 0.1 M Na_2SO_4
- ✓ 0.1 M $Cu(NO_3)_2$
- ✓ 0.1 M $Pb(NO_3)_2$
- ✓ 0.1 M $Ni(NO_3)_2$
- ✓ 0.5 M HCl

PROCEDURE

Part A: Activity Series

1. Obtain small pieces of copper (Cu), magnesium (Mg), and zinc (Zn) metal. In separate test tubes, treat small pieces of each metal with 2–3 mL of the various solutions as outlined below.

   Zn with $MgSO_4$ Cu with $MgSO_4$ *bubbles, cooler*   Mg with $CuSO_4$ *disentegrate, bubbles*

   Zn with $CuSO_4$ *turns black*   Cu with $ZnSO_4$ Mg with $ZnSO_4$

   Zn with HCl *bubbles*   Cu with HCl   Mg with HCl *bubbles - gaseous, cloudy, warmer precip.*

2. After 5 to 10 minutes, observe the color of each solution and examine the surface of each metal for evidence of any deposit. Record your observations in the Data section.

Part B: Precipitation Reactions

1. The 24-well plate is arranged into six columns (numbered 1 through 6) and four rows (lettered A through D). Fill each of the four wells of column 1 half full with sodium carbonate solution. In a similar manner, fill each of the four wells in column 2 with sodium hydroxide solution. Fill each of the wells of column 3 with sodium iodide solution and each of the wells of column 4 with sodium sulfate.

   CAUTION!
   Sodium hydroxide is caustic! Avoid contact with skin, eyes, or clothing.

2. Now in row A, add barium nitrate solution into each of the six wells. In row B, add copper nitrate solution to each of the six wells; in row C, add lead nitrate solution to each of the wells; and in row D, add nickel nitrate to each of the wells. In the Data section, there is a grid where you can record your observations.

3. Once you have completed the above reactions, you can test an unknown anion solution. Use column 6 for your unknown. By mixing your unknown with each of the cations, you should be able to determine its identity.

Catalysis

Catalytic converters in cars are a practical example of catalysis.

Substances that influence the rate of a chemical reaction without undergoing any permanent change themselves are called **catalysts**. The exact mechanism for the action of catalysts is not always completely understood. In general, it is assumed that catalysts provide an alternate path or series of steps by which a reaction can take place. This alternate pathway may bypass some of the slower steps in the uncatalyzed reaction.

There are several examples of catalysis that should be familiar. For example, since 1976, every new car and truck sold in the United States has a "catalytic converter." A catalytic converter consists of finely divided palladium, platinum, or rhodium on a ceramic support. Their purpose is to reduce the emission of the smog-causing compounds—carbon monoxide, nitrogen oxides, and hydrocarbons—by "converting" them into less harmful compounds. By way of illustration, let's look at the conversion of carbon monoxide to carbon dioxide. Normally, this reaction is quite slow in an automobile engine. The catalytic converter increases the rate of this reaction to reduce the emission of toxic carbon monoxide.

$$2\,CO + O_2 \longrightarrow 2\,CO_2$$

One of the major environmental problems facing the world today is due to catalysis occurring in our stratospheric ozone. Stratospheric ozone protects the Earth from dangerous ultraviolet radiation. Since 1970, scientists have noted a dramatic decrease in stratospheric ozone, particularly over the South Pole. The destruction of ozone is catalyzed by chlorine atoms. When a chlorine atom encounters an ozone molecule, the following reactions occur:

The net result is the destruction of one ozone molecule.

$$Cl + O_3 \longrightarrow ClO + O_2$$
$$ClO + O \longrightarrow Cl + O_2$$
$$\text{Net reaction:} \quad O + O_3 \longrightarrow 2\,O_2$$

In the first step of the above reaction sequence, chlorine atoms are reactants, but in the second step they are products (and vice versa for the ClO). Thus, the chlorine atom is regenerated to react with more ozone molecules. A single chlorine atom may be involved in destroying as many as 100,000 ozone molecules before it is carried back to the lower atmosphere. One of the sources of chlorine atoms in the stratosphere comes from the degradation of man-made chlorofluorocarbons (CFCs). In response to the problem of stratospheric ozone depletion, approximately 100 nations have signed the Montreal Protocol, an agreement that aims to reduce and eventually eliminate the emissions of man-made ozone-depleting substances.

P. Crutzen, M. Molina, and F. S. Rowland were awarded the 1995 Nobel prize in chemistry for their work concerning the formation and decomposition of ozone.

In the preceding examples, the catalysts have been relatively simple. In biological systems, however, the catalysts are often very large, complex protein molecules called **enzymes**. Enzymes may have molecular weights that range from several thousand to a million or more.

In this experiment, you will observe the effect of a catalyst by determining the rate at which hydrogen peroxide decomposes in the presence of a catalyst.

$$2 H_2O_2 \xrightarrow{Fe^{3+}} 2 H_2O + O_2$$

The decomposition of hydrogen peroxide to produce water and oxygen gas proceeds slowly at room temperature, but the rate of this reaction can be greatly increased by the presence of iron ions (Fe^{3+}). You will monitor the rate by measuring the volume of water displaced by the oxygen gas formed.

Some reactions are catalyzed by more than a single substance. In biological systems, the ubiquitous enzyme *catalase* also catalyses the decomposition of hydrogen peroxide. Catalase is an enormous protein molecule with a molecular weight of 250,000, but the small portion of it that functions in catalysis appears to involve the four iron ions that are present in each molecule.

MATERIALS

Equipment
- pipet bulb
- 10-mL graduated cylinder
- 100-mL graduated cylinder
- small beaker
- pinch clamp
- oxygen generator
- tubing assembly
- 500-mL Florence flask
- stop watch

Chemicals
- hydrogen peroxide solution, 3%
- 1.0 M ferric chloride solution
- 0.25 M ferric chloride solution
- catalase solution

PROCEDURE

The assembled apparatus is shown below.

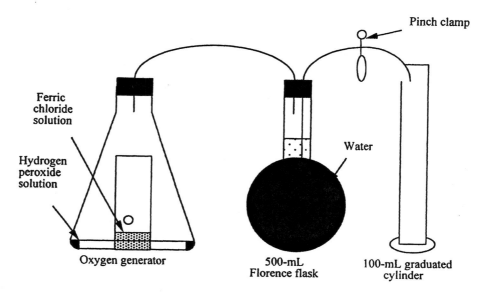

1. Fill a Florence flask with water and attach the tubing assembly so that the long piece of tubing in the flask extends nearly to the bottom and the short piece is above the surface of the water.

2. Place the end of the tubing assembly without the rubber stopper into the 100-mL graduated cylinder.

3. Fill the tubing that connects the Florence flask and the graduated cylinder with water. You can do this by using the pipet bulb to apply pressure from the other end of the tubing assembly (the one with the stopper that fits the Erlenmeyer flask). This pressure forces water from the Florence flask into the graduated cylinder. At the end of this step, the water levels in the Florence flask and the graduated cylinder should be equal and there should be no air bubbles in the connecting tubing.

4. Close the tubing between the Florence flask and the cylinder with the pinch clamp and then discard the water in the cylinder.

5. Obtain about 20 mL of hydrogen peroxide solution in a beaker (enough for all three runs). Use a 10-mL graduated cylinder to measure 5 mL of hydrogen peroxide solution and pour it into the oxygen generator. Make sure that none of this solution gets into the inner tube.

6. Put 0.5 mL of the 1.0-M ferric chloride solution into the inner tube of the oxygen generator.

7. Connect the oxygen generator to the Florence flask by attaching the stoppered end of the tubing assembly.

8. Open the pinch clamp. A few drops of water should flow into the graduated cylinder. A steady stream of water or a continual dripping indicates an air leak. An air leak can be eliminated by making sure all of the stoppers are tight and trying the procedure again. When you are sure there are no air leaks, shake the oxygen generator flask gently (but thoroughly) so that the two solutions mix. The pinch clamp should remain open during this operation. The gas evolved will force an equivalent volume of water into the graduated cylinder.

9. Continue to gently shake the oxygen generator. Record the volume of water collected in the graduated cylinder at 15-second intervals until the volume stops increasing. Record the final volume of water in the graduated cylinder.

10. Repeat the experiment, using 0.5 mL of the 0.25-M ferric chloride solution in place of 1.0-M ferric chloride.

11. Repeat the experiment a third time, using 0.5 mL of catalase solution in place of 1.0-M ferric chloride.

Determination of pH by Using Vegetable Indicators

Many flowers, fruits, and vegetables contain organic compounds that change color with pH.

Hydrangeas will be blue or pink, depending on the acidity of the soil.

Acids and bases were originally classified by their physical properties such as taste: acids tasting sour and bases tasting bitter. It was also noticed that many natural substances contained pigments that changed color when exposed to acids and bases.

In the 1920s, a more systematic definition of acids and bases was proposed by Brønsted and Lowry. They defined an acid as a substance that is capable of donating a proton; conversely, a base is a substance that accepts protons. An example of this is shown below:

$$HCl + H_2O \longrightarrow H_3O^+ + Cl^-$$

In this example, hydrochloric acid, HCl, is the acid and donates a proton (H$^+$) to the base, H$_2$O, water. Hydrochloric acid is defined as a *strong* acid; in other words, it is an excellent proton donor. When hydrochloric acid is mixed with water, the above reaction essentially goes to completion, with all the HCl being converted to H$_3$O$^+$, hydronium ion, and Cl$^-$, chloride ion.

Acetic acid (which is found in vinegar) is said to be a *weak* acid:

$$CH_3COOH + H_2O \longrightarrow CH_3COO^- + H_3O^+$$

A weak acid is not a good proton donor, so when acetic acid is mixed with water, few hydronium ions are formed. If equal molar amounts of acetic acid and hydrochloric acid are put into separate containers of water, the solution of acetic acid has a low concentration of hydronium ions, whereas the solution of hydrochloric acid has a high concentration of hydronium ions.

To measure the strength of an acidic solution, we measure the concentration of H_3O^+. The scale used to denote the acidity of a solution is pH. The pH of a solution is equal to $-\log[H_3O^+]$. Because this log is negative, a high concentration of H_3O^+ corresponds to a low pH. Thus, a 0.1 molar (0.1 M) solution of HCl, a strong acid, has a pH of 1 [$-\log(1 \times 10^{-1})$]. A 0.1 M solution of acetic acid, a weak acid, has a pH of 3 [$(-\log(1 \times 10^{-3})$]. This pH is higher because, of the 0.1 moles of acetic acid present in each liter of solution, only 0.001 moles, or 1/1000, of the total amount is dissociated at any point in time.

$pH = -\log[H_3O^+]$, the brackets indicate that the amount of hydronium ion is stated in molarity (M).

In pure water, there is a slight dissociation of the water molecules into hydronium ions and hydroxide ions:

$$2 H_2O \longrightarrow H_3O^+ + OH^-$$

In pure water, the concentration of hydronium ions is 1×10^{-7} M and because every time a H_3O^+ is formed an OH^- is formed, the concentration of OH^- is also 1×10^{-7} M. Pure water has a pH of 7 [$(-\log(1 \times 10^{-7})$]. Because the concentration of OH^- = H_3O^+, pure water is neither acidic nor basic, but is considered neutral.

Most familiar bases are ionic compounds that contain a hydroxide ion, such as sodium hydroxide (NaOH) or potassium hydroxide (KOH). These bases accept protons from acids to form water:

$$H_3O^+ + OH^- \longrightarrow 2 H_2O$$

Again, a strong base such as NaOH will completely dissociate in water to give a high concentration of hydroxide ions. Some bases do not themselves contain hydroxide ions, but form hydroxide ions when in aqueous solutions through a reaction with water. Ammonia (NH_3) is an example of such a base:

$$NH_3 + H_2O \longrightarrow NH_4^+ + OH^-$$

The pH system is also used to measure the strength of a base. How can this be done when pH is determined by the concentration of hydronium ions but a base usually involves hydroxide ions? Remember that the concentrations of hydronium ion and hydroxide ion in pure water are both equal to 1×10^{-7} M. The product of the concentration of hydronium ion and hydroxide ion is called the ion product of water, K_w.

$$K_w = [H_3O^+][OH^-] = [1 \times 10^{-7}][1 \times 10^{-7}] = 1 \times 10^{-14}$$

The ion product of water is a constant (at constant temperature) and is the same for all aqueous solutions. Because this parameter is constant, when the concentration of hydronium ion increases, the concentration of hydroxide ion must decrease and vice versa. When a base is present, the concentration of hydroxide ion increases and the concentration of hydronium ion decreases. Because the pH is $-\log[H_3O^+]$, when the concentration of hydronium ion decreases, the pH increases.

Just as there are strong and weak acids, there are also strong and weak bases. In a 0.1 M solution of NaOH, the *strong* base is completely dissociated; therefore, the $[OH^-] = 0.1 = 10^{-1}$. Because the product of $[OH^-]$ and $[H_3O^+]$ must be 10^{-14} for all aqueous solutions, $[H_3O^+]$ for this basic solution must equal $10^{-14}/10^{-1}$ or 10^{-13}. The pH of a 0.1 M solution of NaOH thus is $-\log[10^{-13}] = 13$. In a 0.1 M solution of ammonia (NH$_3$), a weak base, only about 1/100 of the ammonium hydroxide is dissociated at any point in time, so $[OH^-] = 0.1/100 = 10^{-3}$. The pH of a 0.1 M ammonia solution is $-\log[10^{-14}/10^{-3}] = -\log[10^{-11}] = 11$.

How does one determine the pH of a solution? One of the earliest methods for determining the pH of a solution was to use chemical compounds that are derived from plants and change color with the pH of the solution. These substances are called **indicators**. One such indicator, litmus, will turn blue in base and red in acid. Some indicators have a wide range of color changes. In this experiment, you will extract colored substances from red cabbage and use them as an indicator. First, you will determine the color the indicator will be at a specific pH, using the buffer solutions of known pH provided for you in the lab. Then you will take various household substances and common laboratory solutions and determine the pH of their aqueous solutions.

MATERIALS

Equipment

- 500-mL beaker
- 100-mL graduated cylinder
- test tubes and rack
- eyedropper
- grease pencil
- hot plate or other heat source
- mortar and pestle

Chemicals

- red cabbage
- buffer solutions
- various household products
- 0.1 M HCl
- 0.1 M acetic acid
- 0.1 M ammonium hydroxide

SAFETY NOTES

In case of spills, wash your skin thoroughly with water and clean up the lab bench.

Use caution with the household chemicals; familiarity tends to breed carelessness.

PROCEDURE

1. Place several purple cabbage leaves in a 500-mL beaker and cover the leaves with water. Boil the cabbage leaves to remove the pigment.

2. While the indicator cools, set up an array of buffer solutions of different pHs. Label 10 small test tubes to correspond with the pHs of the buffer solutions (pHs 2 to 11). Fill each tube approximately one-half full (~5 mL) with the appropriate buffer from the stock solution bottle.

3. Arrange the tubes in order of increasing pH values in your test-tube rack. Add several drops of the cabbage indicator to each tube with an eyedropper. You should get a nice array of colors from the vegetable dye. Record the color of each pH. It's okay to add more of the cabbage indicator to intensify the color but be sure to add the same amount to each test tube.

4. Test the pH of the household products available in the lab by placing a 7-mL sample in a test tube and adding 3mL cabbage indicator. Solid samples such as vitamin C or antacid tablets should be prepared by crushing about one-quarter of the tablet and dissolving it in ~5 mL of deionized water. Compare the resulting colors with your buffer array to determine an approximate pH.

5. With the cabbage indicator, test the pH of several laboratory reagents.

6. Test using pre calibrated pH meters

Experiment 17

Acid-Base Titration

Titration is an analytical method used to determine the concentration of a solution.

Quantitative measurements determine the quantity of each ingredient in a substance. **Qualitative analysis** determines the constituents of a substance without regard to the quantity of each ingredient.

Analytical chemistry is the science of making quantitative measurements. Besides performing reactions, chemists are often called on to determine the identity and/or amount of a chemical in a sample. For example, forensic chemists can be called upon by drug enforcement personnel to determine the identity of an unknown "white powder." There are many analytical methods used by chemists; one of them is titration. **Titration** is the quantitative measurement of a compound in solution done by completely reacting it with a reagent. One type of titration is an acid-base titration.

An acid-base titration utilizes the reaction between an acid and a base is called a **neutralization reaction**. An example of an acid-base reaction is the reaction of an aqueous solution of hydrochloric acid, HCl, with an aqueous solution of sodium hydroxide, NaOH.

$$HCl\ (aq) + NaOH\ (aq) \longrightarrow NaCl\ (aq) + H_2O\ (l)$$

The products of such a neutralization reaction are a salt and water. Because NaOH consists of Na^+ and OH^- ions and the HCl ionizes in water to give H_3O^+ and Cl^- ions, we may rewrite the above equation as

$$H_3O^+ + Cl^- + Na^+ + OH^- \longrightarrow Na^+ + Cl^- + 2 H_2O$$

Ions that appear on both sides of the equation are called **spectator ions**.

If we ignore the ions that appear on both sides of the equation, the resulting equation is

$$H_3O^+ + OH^- \longrightarrow 2 H_2O$$

The balanced chemical equation tells us the number of moles of base required to completely react with a given number of moles of acid. Titration is a common laboratory procedure used to determine the concentration of a solution of base (or acid) by reacting the base with a known amount of acid (or base) of known concentration.

To perform a titration, a known volume of an acid of known concentration is placed in a flask. Then base is added from a buret until just enough has been added to completely react with the acid. This point can be determined by use of an indicator that changes color when the reaction is complete. The point at which the indicator changes color is called the **end point** of the titration.

In this experiment, you will determine the unknown concentration of an acid by using a base of known concentration. The first step in this experiment will be to standardize the base. **Standardization** is the determination of the *exact* concentration of the compound— in this case, the base. We will use a **primary standard**, a solution of known concentration, to determine the concentration of the base. Once the concentration of the base has been determined, it can be used to determine the concentration of an acid. This process may seem complex, but when making quantitative measurements, it is important that the amounts be known exactly.

So how do we determine the concentration of a base by titration? The concentration of the base solution in this experiment can be calculated from the titration data by using the fact that, at the end point, the moles of HCl and NaOH are equal. During the titration, we add acid until the indicator changes color. The volume of acid added contains a number of moles equal to the number of moles of base. If we know the concentration of the acid, we can calculate the molarity (moles/liter) of the base by using the following equation:

(Molarity HCl) × (volume HCl) = (molarity NaOH) × (volume NaOH)

For example, suppose 25 mL of 0.5 *M* HCl were placed in a flask. In other words, there are

25 mL × 0.5 moles/1000 mL = 0.0125 moles HCl

To reach the end point, 0.0125 moles of NaOH will also have to be titrated into the flask. If 12.5 mL (0.0125 liter) of NaOH solution were dispensed from the buret to reach the end point, then the concentration of NaOH would be

0.0125 moles NaOH/0.0125 liter = 1 *M* NaOH

Now let's repeat the calculation using the above equation:

(Molarity HCl) × (volume HCl) = (molarity NaOH) × (volume NaOH)

or

(0.5 *M* HCl) × (25 mL HCl) = (? *M* NaOH) × (12.5 mL NaOH)

If we divide both sides of the equation by 12.5 mL NaOH then

A **buret** is an apparatus for delivering measured quantities of liquid. It consists essentially of a graduated glass tube, usually furnished with a small aperture and stopcock (see below).

$$\frac{0.5 \ M \ HCl \times 25 \ mL \ HCl}{12.5 \ mL \ NaOH} = 1 \ M \ NaOH$$

Now that you have standardized the base, you can use the base to determine the concentration of another acid. The same formula and the same steps are involved in this determination.

This lab is *quantitative*, so careful attention to technique is necessary. The indicator used is thymol blue. It is pink in strongly acidic solutions, yellow in mildly acidic to neutral solutions, and blue in basic solutions. These multiple color changes help to quicken a titration of strong acid/strong base because you can do a fast titration to the yellow color and then a slow titration to the end point. The color of the solution at the end point will be blue for titration of the acid. At the *exact* end point the solution is neutral; you have added an amount of basic solution that contains the number of moles equivalent to the moles of acid. At one drop past the end point the base is in excess, and this excess will make the solution basic. With the intermediate yellow color, you have a warning that the end point is approaching.

MATERIALS

Special Note

A buret is used to perform the titration. Burets are fragile and expensive, so take care.

Equipment
- 100-mL graduated cylinder
- small beakers
- 50-mL Erlenmeyer flasks (2)
- 10-mL pipet
- pipet bulb
- 10-mL buret

Chemicals
- NaOH solution (~0.5 M to 1.0 M)
- standardized HCl solution
- thymol blue indicator solution
- unknown acid solution

PROCEDURE

Part A: Standardization of the NaOH solution

1. Use a buret clamp to secure the 10-mL buret to a ring stand. Thoroughly rinse the 10-mL buret with deionized water. The deionized water should drain evenly from the inside surfaces of the buret and leave no droplets of water behind. Any droplet formation on the inside of the buret means that the buret is dirty and needs to be cleaned.

2. Obtain about 50 mL of sodium hydroxide solution (~0.5 M to 1.0 M) in a small beaker.

3. Fill the buret with sodium hydroxide solution. Open the stopcock long enough to fill the tip with solution and remove any air bubbles. Add solution to the top of the buret, or drain the solution through the stopcock, to bring the level close to 0.0 mL on the buret scale. It is not necessary that the initial level is exactly 0.0 mL, but it must be at or below the start of the scale.

4. Record the initial level of the buret, keeping your eye level with the bottom of the meniscus. Estimate the last digit.

5. Obtain approximately 25 mL of the standardized HCl solution. Record the exact concentration of the HCl.

6. Rinse the 10-mL pipet with two small portions of the acid solution, and then use the pipet to accurately transfer 10-mL portions of this solution into each of two clean 50-mL Erlenmeyer flasks.

7. Add several drops of thymol blue indicator to each flask.

8. Titrate the acid solution to the end point, the first permanent appearance of blue in the solution. Add the base rapidly at first as you continuously swirl the flask and its contents to ensure rapid mixing. As you approach the end point, continue the addition of base more slowly— eventually drop by drop— mixing carefully after each addition. With practice, it is possible to determine the end point with the precision of one drop.

9. Record the final volume in the buret.

10. Repeat the titration on the other 10-mL acid sample. Check the volume of sodium hydroxide that you have left in the buret. You may wish to refill the buret with more solution before beginning the second titration.

11. If the volumes of NaOH delivered for the two trials are not within reasonable agreement, do a third titration.

Part B: Determining the molarity of an unknown acid solution

1. In a clean beaker, obtain at least 25 mL of ~~HCl~~ Vinegar solution of unknown concentration. *Record the number of the unknown you chose.*

2. Rinse the pipet twice with small portions of the ~~acid~~ vinegar solution and then accurately transfer ~~10~~ mL of the acid solution into each of two clean ~~50~~-mL Erlenmeyer flasks. 1.00   250

3. Add 2 drops of ~~thymol~~ phenothalene blue indicator to each flask and titrate each to the end point. Record the initial and final volumes of each titration.

light pink color

DATA

Part A: Standardization of the NaOH solution

Concentration of standard HCl solution: _____

Titration 1

final buret volume _____

initial buret volume _____

volume NaOH _____

Titration 2

final buret volume _____

initial buret volume _____

volume NaOH _____

Part B: Determining the molarity of an unknown acid solution

Unknown number: _____

Titration 1

final buret volume _____

initial buret volume _____

volume NaOH _____

Titration 2

final buret volume _____

initial buret volume _____

volume NaOH _____

RESULTS

1. For the standardization of base, calculate the exact concentration of the NaOH solution using the data from each run. (*Hint:* Use the $M_a V_a = M_b V_b$ relationship, where a = acid and b = base)

Titration 1, concentration of NaOH: _____

Titration 2, concentration of NaOH: _____

2. For the titration of your unknown, calculate the molarity of the unknown concentration of HCl solution using the data from each run

Titration 1, concentration of HCl: _____

Titration 2, concentration of HCl: _____

Experiment 18

Simple Distillation of Organic Solvents

Distillation can be used to purify a solvent or to determine the boiling point of an unknown.

There are a variety of important physical properties that can be explored. One of these is the boiling point of a substance. **Physical properties** are the properties that can be measured without chemically changing the compound. The **boiling point** of a substance is the temperature at which a liquid is converted into vapor. Boiling points are different for different liquids, and for the same liquid under different pressures. Because the boiling point is different for different compounds, it can be used to determine the identity of an unknown liquid.

When you look at different organic functional groups, you will see patterns associated with boiling points. For a compound to vaporize (boil), the attractive forces between the molecules must be broken. The stronger the forces between molecules, the more energy required to change a liquid to a vapor. Therefore, if you understand factors that increase or decrease the attractive forces between molecules, then you can predict general boiling point trends.

There are two general trends observed for organic molecules. Within a group of organic compounds—for example, alkanes—you will see an increase in the boiling point with increasing molecular mass. As the size (molecular mass) of a molecule increases the attractive forces between the molecules increase.

The second trend is that, for molecules of essentially equal molecular mass, the molecule with the stronger attractive forces will have a higher boiling point. There are three basic secondary (attractive) forces. **London forces** are attractive forces resulting from a temporary dipole in the molecule; these are the weakest secondary forces. London forces are seen between nonpolar compounds, such as, alkanes and cycloalkanes. **Dipole-dipole interactions** are attractive forces between polar molecules. **Hydrogen bonding** is a special case of dipole-dipole attraction that is stronger than any other dipolar force. A hydrogen bond has two requirements: hydrogen attached to either oxygen, nitrogen, or fluorine and another oxygen, nitrogen or fluorine. An example of a hydrogen bond is shown below, the hydrogen bond (H-bond) is represented by a dotted or dashed line.

$$H\text{-}F \cdots \cdots H\text{-}F$$

In addition to the determination of the boiling point of a substance, distillation can be used to separate two compounds with different boiling points. Distillation is often used to separate and/or purify organic compounds. Crude oil is separated into various fractions on the basis of different boiling points. For example, gasoline has a lower boiling point than heating oil, so crude oil can be fractionated into a variety of oil-based products by distillation.

In this experiment, you will use distillation to separate two volatile liquids. You will use the boiling points to determine the identity of the individual components. When distillation begins the temperature at the distillation head will be that of the lower boiling component. As the mixture becomes depleted in the low boiling component, the temperature will rise and eventually reach that of the higher boiling component.

Compound	Boiling point
acetone	57°C
n-hexane	69°C
ethyl acetate	77°C
heptane	98°C
toluene	110°C

To return to secondary forces, notice that there are two alkanes in the list, n-hexane and heptane. Heptane has a higher molecular mass and a higher boiling point than hexane, just as one would predict. So you see, with an understanding of the secondary forces that occur between various types of organic compounds, one can predict relative boiling points.

MATERIALS

Equipment

- ✓ distillation apparatus
- ✓ thermometer
- ✓ heat source
- ✓ 100-mL graduated cylinder
- ✓ boiling stones

Chemicals

- ✓ unknown organic mixtures

PROCEDURE

1. Set up distillation apparatus as shown below.

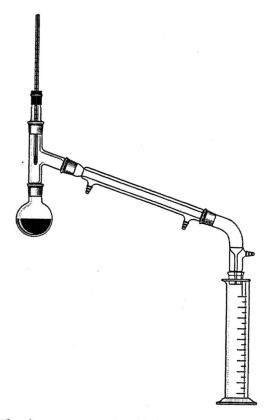

Special Note
To prevent breakage, use clamps to support the assembly.

2. The heat source should be placed on a blocks to permit quick removal in an emergency. Several blocks may be needed to raise the assembly high enough to accommodate the graduated cylinder that is used as a receiver.

3. You must use an appropriate heat source. Do not use a flat hot plate to heat a round-bottom flask. You must use a heat source that will conform to the shape of the round bottom. This can be an electrical thermowell heater or a sand bath. Consult your laboratory assistant to determine your heat source.

4. A clamp should be used as the main support of the assembly and it must be able to hold the weight of the flask (and column) when or if the heater is removed.

5. Obtain 30 mL of an unknown and record the number of your unknown. Place the unknown and 3 to 4 boiling stones in a 100-mL round-bottom flask of a simple distillation assembly. Put the boiling chips into the *cold* mixture to promote smooth boiling (prevents "bumping").

CAUTION!
Never put a boiling chip into hot liquid!

6. Accurate placement of the thermometer bulb is essential to get the temperature of the liquid-vapor equilibrium. It should be below the ring of condensing hot vapor, but not so low that the drop of liquid on the bulb evaporates.

7. Have your laboratory assistant inspect your apparatus before you begin your distillation.

8. Adjust the heating rate so that the mixture distills slowly, approximately 1 drop every 2–3 seconds. Record the temperature at the distilling head at intervals corresponding to every milliliter of distillate collected.

9. As the distillation proceeds, it may be necessary to increase the heat. Stop the distillation when 2–3 mL of liquid remains in the distilling flask or when the distilling head temperature rises to a constant value. *Never* boil a flask to dryness.

Note: There will be a small amount of the purple dye in this mixture; this high boiling compound will not distill from this solution.

DATA

Unknown number: _____

mL distillate	temperature	mL distillate	temperature
_____	_____	_____	_____
_____	_____	_____	_____
_____	_____	_____	_____
_____	_____	_____	_____
_____	_____	_____	_____
_____	_____	_____	_____
_____	_____	_____	_____
_____	_____	_____	_____
_____	_____	_____	_____
_____	_____	_____	_____
_____	_____	_____	_____
_____	_____	_____	_____
_____	_____	_____	_____
_____	_____	_____	_____

RESULTS

1. Plot a graph of temperature versus the volume distilled. You should let the temperature level off at the boiling points of the two components.

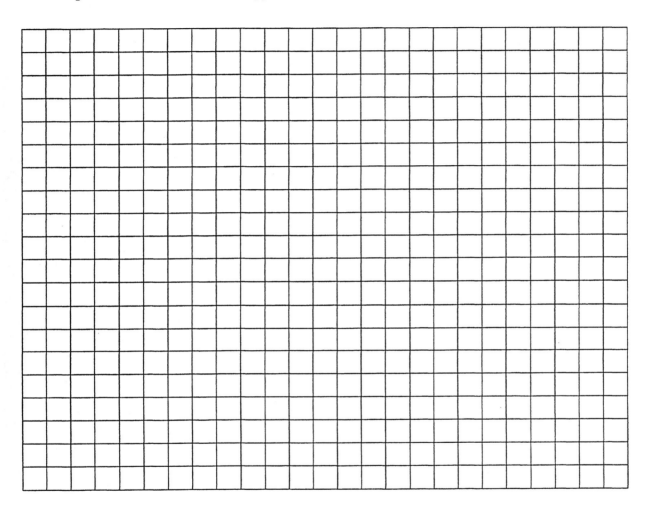

2. From the graph, estimate the boiling points of both components in your unknown. Use the boiling points to determine the identity of the unknowns.

   Boiling point of Component 1: _____   Identity _____

   Boiling point of Component 2: _____   Identity _____

3. What is the percent composition of each component in the mixture?

Physical and Chemical Properties of Saturated and Unsaturated Hydrocarbons

Petroleum is the major source of hydrocarbons.

Saturated and Unsaturated Hydrocarbons

Hydrocarbons, as their name implies, are organic compounds that contain only carbon and hydrogen. They are extremely important to our society because many of the products we depend on are derived from them: fabrics, plastics, antifreezes, gasoline, anesthetics, and insecticides, to name a few.

Aliphatic compounds are non-aromatic compounds, that is, alkanes, alkenes, and alkynes.

Petroleum, which is an extremely complex mixture of compounds, contains mainly aliphatic hydrocarbons of five-carbon molecules and larger. The average U.S. citizen directly or indirectly uses several tons of petroleum each year, mostly as fuel (gasoline, diesel, jet fuel, and heating oil). Because petroleum in a nonrenewable resource, we need to seriously consider our use of these products, particularly our use of fuels.

Two possible representations for benzene

In this experiment, you will observe the physical and chemical properties of three different types of hydrocarbons: alkanes, alkenes, and aryls. Saturated hydrocarbons, **alkanes**, are those that consist entirely of carbon-carbon single bonds. **Alkenes** are unsaturated hydrocarbons and contain carbon-carbon double bonds. **Aryls** compounds are also unsaturated hydrocarbons but are generally referred to as aromatic compounds. The simplest and most common example of an aromatic compound is benzene. Organic compounds that contain a benzene ring are called **aromatic compounds.**

Even though there are literally millions of organic compounds, the study of organic chemistry is made easier by the fact that groups of organic compounds demonstrate the same behaviors. We divide organic compounds into families that have a common structural feature called a **functional group**. Each member of a family will undergo similar chemical reactions; therefore, it is not necessary to remember the reactions of millions of compounds. One need only understand the reactivity of particular functional groups and this information can be applied to compounds that belong to that family. The table below lists hydrocarbon family names, the functional group, and compounds that are examples of each type.

Family	Characteristic functional group	Examples
alkanes	contain carbon-carbon single bonds	methane, CH_4 cyclohexane, C_6H_{12}
alkenes	contain carbon-carbon double bonds	ethene, C_2H_4 cyclohexene, C_6H_{10}
aryls	contains a benzene ring*	benzene, C_6H_6 toluene, C_7H_8

* Molecules with benzene rings are just one example of aromatic (aryl) compounds.

In this experiment, you will look at the physical and chemical properties of these families. **Physical properties** are the properties that can be measured without chemically changing the compound. Examples of physical properties are density, solubility, boiling point, and melting point. A **chemical property** describes how a compound reacts to form new compounds. Each of these types of compounds undergoes distinctive reactions. Reactions that can be used to distinguish between these three types of hydrocarbons are given below.

Reactions with Halogens

Each of these hydrocarbon types undergoes a different reaction with halogens. Bromine, for example, reacts differently with each of these hydrocarbons. Alkanes require sunlight to initiate the reaction, which often takes a considerable length of time to occur. Alkenes are very reactive and react immediately with bromine. Aryl compounds are unreactive and will only react with bromine or other halogens in the presence of a catalyst. Each of these different reactions is illustrated below in a general form, using the letter R to represent an alkyl group.

The halogens are
fluorine (F_2)
chlorine (Cl_2)
bromine (Br_2)
iodine (I_2)

Alkanes

$$R-CH_3 + Br_2 \xrightarrow{\text{Sunlight}} RCH_2Br + HBr$$

Alkane — Reddish brown — Colorless — Gas

Alkenes

$$\underset{\underset{R}{R}}{\overset{\overset{R}{|}}{C}}=\underset{\underset{R}{|}}{\overset{R}{C}} + Br_2 \longrightarrow R-\underset{\underset{Br}{|}}{\overset{\overset{R}{|}}{C}}-\underset{\underset{Br}{|}}{\overset{\overset{R}{|}}{C}}-R$$

Alkene   Reddish brown   Colorless

Aryls

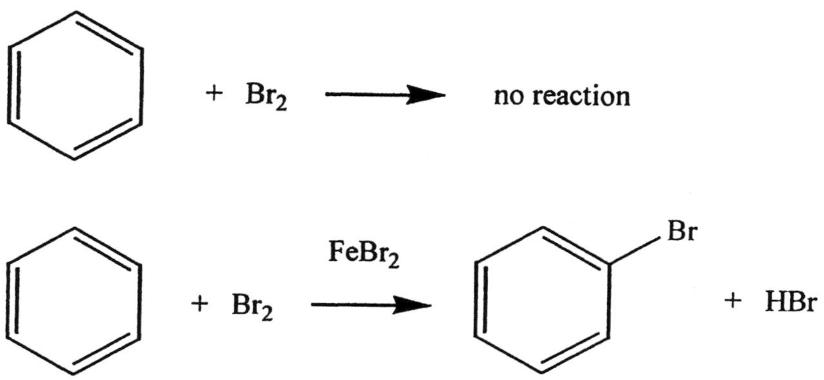

benzene + $Br_2 \longrightarrow$ no reaction

benzene + $Br_2 \xrightarrow{FeBr_2}$ bromobenzene + HBr

Reactions with Potassium Permanganate

Potassium permanganate, $KMnO_4$, reacts with alkenes but not with alkanes and aryls. Evidence of this reaction is the disappearance of purple $KMnO_4$ and the formation of brown manganese dioxide, MnO_2.

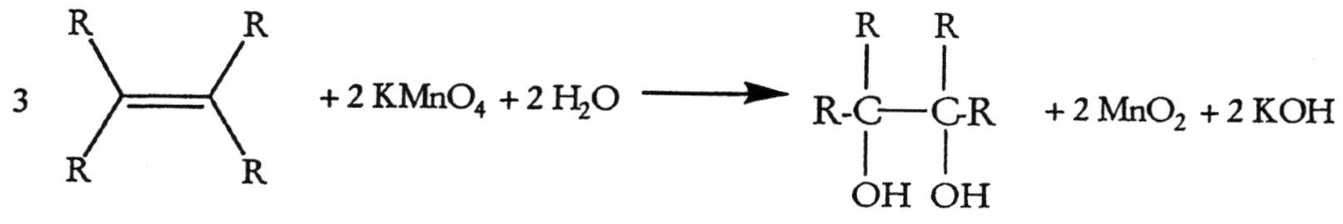

$$3\ \underset{\underset{R}{R}}{\overset{\overset{R}{}}{C}}=\overset{R}{C} + 2\ KMnO_4 + 2\ H_2O \longrightarrow R-\underset{\underset{OH}{|}}{\overset{\overset{R}{|}}{C}}-\underset{\underset{OH}{|}}{\overset{\overset{R}{|}}{C}}-R + 2\ MnO_2 + 2\ KOH$$

Alkene   Purple   A glycol   Brown

Solubility of Hydrocarbons

In addition to chemical properties, you will look at the physical properties of these hydrocarbons. Unlike their chemical properties, which are different, the physical properties of these three groups are similar. You will look at the solubility of these compounds with respect to a polar solvent, water, and a nonpolar solvent, methylene chloride (CH_2Cl_2). As you learned in general chemistry, "like dissolve likes." Using these two different solvents, you will be able to determine whether the tested hydrocarbon is a polar or a nonpolar compound.

Relative density can be determined by examining the density of a compound with respect to water. If the hydrocarbon is denser than water, it will sit on the bottom of the test tube; if it is less dense than water, it will sit as a layer on top of water.

The density of water is 1 gram/mL.

MATERIALS

Equipment
- test tubes
- test-tube rack
- eyedropper
- sunlamp (optional)

Chemicals
- bromine solution
- 2% $KMnO_4$ solution
- cyclohexane
- cyclohexene
- toluene
- methylene chloride
- ethanol

CAUTION!
Bromine vapors can be irritating. Do all bromine additions in a hood.

PROCEDURE

Part A: Exploring physical properties

1. Place about 10 drops of cyclohexane, cyclohexene, and toluene in three separate, clean, dry test tubes. Compare the odors of the three hydrocarbons. Remember to waft the vapors to your nose from 6 to 12 inches away.

   Waft: To carry the vapors to your nose using a cupped hand.

2. Add about 10 drops of water to each test tube. Shake each test tube and note solubility. Also, note whether the hydrocarbon is more or less dense than water. If you are unable to tell which layer is the hydrocarbon, add a few more drops of water and see which layer gets larger— that's the water layer.

   If the hydrocarbon is insoluble, you should see two layers in the test tube.

3. Repeat the solubility test, using methylene chloride, CH_2Cl_2, and fresh samples of the hydrocarbons in clean test tubes.

4. Report your observations below and answer the questions.

Part B: Exploring chemical properties

1. Place 10 drops of each hydrocarbon in separate clean, dry test tubes.

2. Add 2 drops of dilute bromine in methylene chloride to each of the test tubes. Observe whether a reaction (bromination) occurs.

3. For any sample that does not react, place that test tube under a sunlamp or next to a sunny window, if available. Leave the test tube there for the remainder of the period. At the end of the lab period, check and see whether any reaction has occurred.

4. To test the reactivity with permanganate, in separate test tubes dissolve 6 drops of each hydrocarbon in 2 mL of ethanol (a solvent).

5. Add 2 drops of 2% $KMnO_4$ solution. A positive test is the replacement of the purple color of the $KMnO_4$ solution by a brown precipitate. The reaction should occur in one minute or less. The solvent, ethyl alcohol, also reacts with $KMnO_4$ but the reaction is slower, producing a brown color in about 5 minutes. Do not mistake this slower reaction for a positive reaction.

6. Report your observations below and answer the questions.

NAME_____

RESULTS

Part A: Physical properties

Compound	Soluble in H₂O (yes/no)	Soluble in CH₂Cl₂ (yes/no)	Density (more/less than water)
cyclohexane			
cyclohexene			
toluene			

Describe any difference you detect between the odor of the three compounds.

Part B: Chemical properties (Include color changes in your observations)

Compound	Reacts with Br₂ (yes/no)	Reacts with Br₂ + sunlight (yes/no)	Reacts with KMnO₄ (yes/no)
cyclohexane			
cyclohexene			
toluene			

QUESTIONS

1. Write chemical equations to illustrate the reaction of bromine with cyclohexane and cyclohexene.

2. Which is more reactive, an alkane or an alkene? Explain in your own words why one of these families is more reactive than the other.

3. You may have seen some reaction of bromine with toluene. Because a catalyst is not present, the benzene ring in toluene should not react. Look at the structure of toluene. Where do you think the reaction occurred? Is this reaction similar to the reaction with an alkane or an alkene? Explain.

   Toluene

4. What type of reaction is occurring between potassium permanganate and cyclohexene?

Molecular Shape and Isomerism in Organic Chemistry

Molecular shape is an important factor in the properties of organic compounds.

The purpose of this exercise is to help you visualize the three-dimensional structure of organic molecules. In this exercise, you will study two types of isomers: constitutional isomers and geometric isomers.

Constitutional (structural) isomers are two or more molecules that have identical molecular formulas but different arrangements of their atoms. We find so many different organic molecules, because carbon compounds can form long chains of carbon with one or more carbon branches. An example of a pair of constitutional isomers with the molecular formula C_4H_{10} is shown below:

$$CH_3\text{-}CH_2\text{-}CH_2\text{-}CH_3 \qquad\qquad CH_3\text{-}CH\text{-}CH_3$$
$$\qquad\qquad\qquad\qquad\qquad\qquad\qquad |$$
$$\qquad\qquad\qquad\qquad\qquad\qquad\qquad CH_3$$

            n-Butane   2-Methylpropane

Constitutional isomers are also possible for alkenes. With alkenes, there can exist different carbon skeletons (like those shown above) and also different placements of the double bond. Below are examples of compounds with the same carbon skeleton but different double bond placement.

$$H_3CHC=CHCH_3 \qquad\qquad H_2C=CHCH_2CH_3$$

            2-Butene   1-Butene

We see another type of isomerism when we have restricted rotation around a bond. We see such restricted rotation in alkenes and cycloalkanes. The restricted rotation results in geometric isomerism also known as cis-trans isomerism or stereoisomerism. Geometric

isomers are molecules that have the same molecular formula and the same order of attachment of atoms, but a different orientation of their atoms in space. The constitutional isomer 2-butene can serve as an example of cis-trans isomerism:

cis-2-Butene   *trans*-2-Butene

A cis isomer has the two substituents— in this case, methyl groups— on the same side of the double bond. Trans isomers have the substituents on opposite sides of the double bond.

> A **substituent** is any atom (except hydrogen) or group of atoms that branch off from a carbon.

These different types of isomerism are not just a curiosity. They are important in determining the different properties of organic compounds. For example, when comparing constitutional isomers of alkanes, branching results in a decrease in the boiling point. Branched alkanes are more compact in shape and have a lower surface area. Contact over a larger surface area results in greater secondary (intermolecular) forces between molecules, which in turn leads to a higher boiling point than that of branched alkanes with similar mass. This is just one of example of the effect of isomerism.

In this exercise, you will build a variety of organic molecules and explore isomerism. To do this, you will use a model kit. The molecular model kit contains plastic balls and sticks to represent atoms and bonds. The black balls with four holes represent carbons; the white balls represent hydrogen; and the colored balls represent other atoms, such as nitrogen, oxygen, or the halogens. Connectors are used for bond formation, and those for single bonds and double bonds generally look different. Check the kit and be sure to use the correct connectors.

You will also look at addition reactions of alkenes and explore the types of products that can be formed.

MATERIALS

Equipment
✓ molecular model kit

EXERCISE

Do each of the following sections. For each section, draw the molecule you have constructed and answer the questions.

Part A: Alkanes

1. Construct a model of methane (CH_4). Draw a diagram of methane below. Use wedges to indicate a bond that comes out of the plane of the paper and a dashed line to indicate a bond that go behind the plane of the paper. Compare your drawing with the model to familiarize yourself with the two-dimensional drawing as a representation of a three-dimensional object.

2. Remove one of the hydrogens. The remaining group of atoms is a methyl group. How would you represent a methyl group to show how it differs from methane?

Part B: Constitutional Isomers

1. Construct a model of *n*-pentane (C_5H_{12}). Rearrange the carbon atoms to form all the possible isomers of *n*-pentane. Verify that all the isomers are different. Draw all the constitutional isomers of C_5H_{12} below. Name each of the compounds, using IUPAC nomenclature.

2. Which of the isomers that you constructed appears to be the most compact?

Part C: Geometric isomers of alkenes

1. Construct a model of ethene (C_2H_4). Next, replace two of the hydrogens with chlorine (use green balls if you have them) such that you form the cis isomer. Draw and name this compound below.

2. Construct the trans isomer of the above compound. Draw and name this compound below.

3. Is there any way to convert one isomer to another without breaking a bond?

4. Using the same set of atoms ($C_2H_2Cl_2$) construct a molecule that is a *constitutional* isomer of the cis-trans isomers constructed above. Draw and name this compound below.

Part D: Geometric and stereoisomers of cycloalkanes

1. Construct a cyclobutane molecule (C₄H₈). Replace two hydrogens with chlorines to make a cis-dichloro isomer of cyclobutane. Draw and name this compound.

2. How many cis-dichlorocyclobutanes can you construct? Draw and name them.

3. Convert the cis to the trans isomer for each of the above. Draw and name.

Part E: Alkene addition reactions

1. Construct a propene molecule (C₃H₆). Using two same-colored atoms to represent chlorine, break the double bond and add the atoms. Write the reaction and products.

2. Convert the above molecule back to propene and make a second propene molecule. Construct two molecules of HCl. Add an HCl molecule to each of the propene molecules to form two different chloropropane molecules. Write the chemical equation; draw and name the products below.

3. Which of the above products is the preferred Markovnikov product?

Experiment 21

Organic Qualitative Analysis

Qualitative analysis is used to identify unknown compounds.

Organic qualitative analysis is a process by which one performs a series of tests on an unknown compound in order to determine the identity of that compound. In this experiment, you will use four simple organic qualitative tests to identify unknown compounds. The unknown compounds could be an alcohol, a phenol, an aldehyde, or a ketone. Knowing the structure of possible unknowns, you can predict how they will react in each test. These results will allow you to determine the identity of each unknown.

The four tests you will perform are the dichromate test, the DNPH (2,4-dinitrophenylhydrazine) test, the iodoform test, and the ferric chloride test. In each of these tests, a positive reaction denotes the presence of a particular structure or functional group. Each test is discussed below.

Dichromate Test

Potassium dichromate (K_2CrO_7) is an oxidizing reagent that will oxidize primary and secondary alcohols. In a dichromate test, primary and secondary alcohols are oxidized to aldehydes and ketones, respectively. Dichromate solution will also oxidize aldehydes to carboxylic acids. The test will be negative for ketones and tertiary alcohols because they are inert to further oxidation under these mild conditions.

Primary alcohol

$$R\text{-}CH_2OH \xrightarrow[H_2SO_4/H_2O]{K_2CrO_7} \underset{R}{\overset{O}{\underset{\|}{C}}}\text{-}H \xrightarrow{[O]} \underset{R}{\overset{O}{\underset{\|}{C}}}\text{-}OH$$

Secondary alcohol

$$R_2CHOH \xrightarrow[H_2SO_4/H_2O]{K_2CrO_7} \underset{R}{\overset{O}{\underset{\|}{C}}}\text{-}R$$

DNPH Test

The 2,4-dinitrophenylhydrazine (DNPH) test is used to identify aldehydes and ketones. DNPH reacts with the carbonyl of an aldehyde or a ketone to form a yellow-orange solid that precipitates out of solution.

2,4-Dinitrophenylhydrazine (DNPH)   Yellow-orange precipitate

Iodoform Test

The iodoform test is used for the identification of the following two groups:

$$R-\overset{O}{\underset{\|}{C}}-CH_3 \quad \text{and} \quad R-\overset{OH}{\underset{|}{CH}}-CH_3$$

Compounds containing either of these groups react with iodine in sodium hydroxide to give a bright yellow precipitate of iodoform (CHI_3). Phenols will also react under these conditions to give a yellow precipitate.

Ferric Chloride Test

The ferric chloride ($FeCl_3$) test is used to identify phenols. Light yellow aqueous ferric chloride reacts with phenols to give a purple-colored solution.

Colored complex

There is a pre-lab assignment for this experiment. Completion of the assignment before going to lab will aid you in the identification of your unknowns.

MATERIALS

Equipment

- ✓ test tubes
- ✓ test-tube rack
- ✓ 250-mL beaker
- ✓ 10-mL graduated cylinder
- ✓ dispensing pipets

Chemicals

- ✓ acetone
- ✓ 1-propanol
- ✓ vanillin
- ✓ dichromate reagent
- ✓ iodine reagent
- ✓ DNPH reagent
- ✓ 10% NaOH(aq)
- ✓ ferric chloride(aq)
- ✓ unknowns

> **CAUTION!**
> The nature of these reagents makes it imperative that you wear your safety goggles at all times. Dispose of all waste solutions into the appropriate waste containers.

PROCEDURE

Carry out each of the tests below on the three known compounds: acetone, 1-propanol, and vanillin. When test results agree with the predicted results for these compounds (see pre-lab assignment), then proceed to carry out the four tests on *two* of the unknown compounds, A–F. Also, carry out the four tests on compound X, which has a molecular formula of $C_4H_nO_2$.

Record the results of all tests in the data table.

Dichromate test

1. Add 4 drops of the unknown to about 1 mL of dichromate reagent in a test tube. Mix well.

2. A positive dichromate test is observed as a bluish green solution that gives off heat. A negative dichromate test shows the original orange color of the reagent.

DNPH test

1. Place about 1 mL of DNPH reagent in a test tube. Add 1 drop of the unknown to be tested.

2. A positive DNPH test is observed as the formation of a precipitate. A negative DNPH test shows an unchanged yellow solution.

> Be sure to collect the waste for *each* test in a labeled beaker and at the end of the period dispose of it in the proper waste container.

Iodoform test

1. Place about 1 mL of iodine reagent in a test tube. Add 4 drops of the unknown to be tested. Mix well.

2. Add 10% NaOH solution to the test tube drop by drop with continual mixing until the dark color of the iodine has been removed. A pale yellow color may remain.

3. A positive iodoform test is observed as a heavy yellow to creamy colored suspension from which a precipitate will settle. No evidence of a suspension or precipitate is observed in a negative test.

Ferric chloride test

1. Place about 1 mL of aqueous ferric chloride solution in a test tube and add 5 to 10 drops of your unknown to be tested. Note any color changes in the solution.

2. A positive ferric chloride test can give a color change ranging from green to purple, depending on the structure of the phenol. No color change from light yellow is observed in a negative test.

PRE-LAB ASSIGNMENT

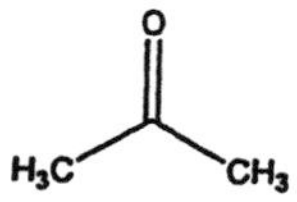

1. For each test compound, predict the result, positive (+) or negative (−), for each of the four tests.

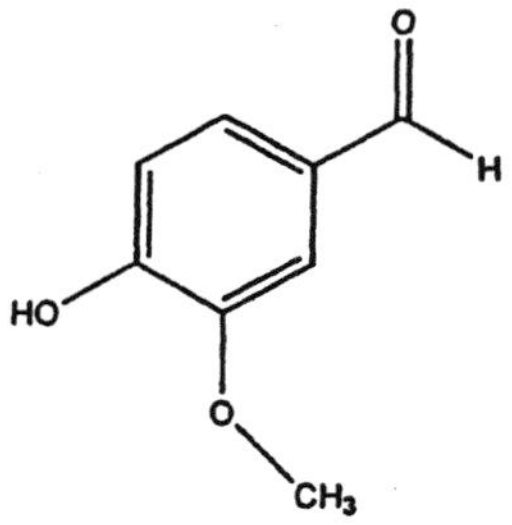

Compound	Dichromate test	DNPH test	Iodoform test	Ferric chloride test
acetone				
vanillin				
1-propanol				

2. The possible unknowns (except X) are listed below. Write the structural formulas for each of the possible unknowns and classify each unknown as an aldehyde, ketone, primary alcohol, secondary alcohol, tertiary alcohol, or phenol.

Possible unknown   **Structure**   **Classification**

propanal

1-butanol

2-butanol

t-butyl alcohol

phenol

2-butanone

NAME_____

DATA

Indicate the letter of the unknown and the results of each test (positive or negative).

Unknown	Dichromate test	DNPH test	Iodoform test	Ferric chloride test
X				

RESULTS

1. Once your results have been recorded, identify the unknowns you tested. Be sure to include your reasoning for your unknown identification. Unknowns A–F could be propanal, 1-butanol, 2-butanol, t-butyl alcohol, 2-butanone, and phenol.

Unknown: _____   **Identity:** _____

State your reasoning for your choice of the unknown.

Unknown: _____   **Identity:** _____

State your reasoning for your choice of the unknown.

2. Compound X has a molecular formula of $C_4H_{10}O_2$. From the results what are the possible functional groups in compound X?

3. Draw and name compound X.

Differential Extraction of Organic Acids

Extraction is a method commonly used to purify a compound.

Carboxylic acids, with more than five carbons, and phenols are not very soluble in neutral water. Phenols, like carboxylic acids, are acidic. These acids can be converted into ionic compounds by reacting them with base. The ionic salts of carboxylic acids and phenols have an increased solubility in water. The effect can be quite dramatic. For example, benzoic acid has a solubility in water of 0.34 g/100 mL. If it is converted to the sodium salt, sodium benzoate, the solubility increases to 50 g/100 mL.

How can we use this solubility difference to separate two acids? If we have a mixture of both a phenol and a carboxylic acid, converting them both to a soluble salt does not help us separate them— now we have them both in solution. We need to use their relative acidities to our advantage in the separation. Phenols are generally less acidic than carboxylic acids and require a stronger base to form the ionic salt. Because of this difference, we can use a weaker base (for example, sodium bicarbonate), which will react with carboxylic acids but not phenols. In this way, we can selectively separate these two different types of acids.

The reaction of a carboxylic acid with sodium bicarbonate is given below:

$$RCO_2H + NaHCO_3 \longrightarrow RCO_2^- Na^+ + CO_2 + H_2O$$

Water-insoluble carboxylic acid   Water-soluble carboxylate salt

Notice that carbon dioxide gas is formed in this reaction. This product will become important later during the procedure, because you must guard against a buildup of gas in the separatory funnel.

Under the above conditions, a phenol will not react. A stronger base (such as sodium hydroxide) is required to form a water-soluble phenoxide ion.

[phenol] —OH + NaOH ⟶ [phenoxide] —O⁻Na⁺ + H₂O

Water-insoluble phenol   Water-soluble phenoxide ion

This difference in acidity will be used to separate these compounds. In this separation, you will first use NaHCO₃ to form the carboxylate salt that is soluble with water. Once the carboxylic acid is removed, a stronger base, NaOH, can be used to react with the phenol to form a water-soluble compound. Once the two are separated into their separate aqueous solutions, the individual salt solutions can be acidified by HCl and converted to the original carboxylic acid and phenol, which are not soluble in water. They will now precipitate from solution.

Reaction of the salts with acid:

$$RCO_2^- \; Na^+ + HCl(aq) \longrightarrow RCOOH + NaCl(aq)$$

[PhO⁻Na⁺] + HCl(aq) ⟶ [PhOH] + NaCl(aq)

By taking advantage of these solubility properties, it is possible to separate a mixture containing a carboxylic acid, a neutral compound, and a phenol. In this experiment, a mixture containing thymol (a phenol), cholesterol (a neutral alcohol), and benzoic acid will be separated, using base extractions.

Benzoic acid   Thymol   Cholesterol

In the pre-lab assignment, you will fill in a flowchart for your separation. This preparation will help you to complete the lab efficiently.

MATERIALS

Equipment

- ✓ separatory funnel
- ✓ ring stand
- ✓ Erlenmeyer flasks (2)
- ✓ small beaker
- ✓ ice bath container
- ✓ Büchner funnel
- ✓ side-arm vacuum flask
- ✓ pH paper
- ✓ test tube
- ✓ glass stirring rod

Chemicals

- ✓ organic mixtures
- ✓ diethyl ether
- ✓ 5% aqueous $NaHCO_3$
- ✓ 1 M aqueous NaOH
- ✓ 6 M HCl
- ✓ saturated NaCl solution
- ✓ anhydrous Na_2SO_4

PROCEDURE

Isolation of the carboxylic acid

Separatory funnel

CAUTION!
No open flames in lab because ether is flammable!

1. Dissolve approximately 1 g of your mixture in 15 mL of diethyl ether. Record the exact mass used.

2. Transfer the ether solution to a separatory funnel (see figure) and add 5 mL of 5% $NaHCO_3$ (sodium bicarbonate solution).

3. Gently shake the separatory funnel for several minutes, venting it *frequently* to avoid the buildup of pressure. Carbon dioxide gas is produced, and heat from your hand is sufficient to vaporize ether.

4. Place the separatory funnel (uncapped) in a ring stand and allow the ether and aqueous layers to separate. Drain the lower aqueous layer into a labeled Erlenmeyer flask. You will be adding another aqueous extract to this flask in Step 6, so don't throw it away!

5. Repeat the extraction with another 5 mL of sodium bicarbonate solution. Save the ether layer in the separatory funnel for use later.

6. Combine the aqueous extracts and cool in an ice bath.

7. Once cooled, add 6 M HCl (≈ 2 mL) until the aqueous layer is acidic. Confirm acidity with pH paper. (Proper technique is to dip a stirring rod into the solution and then touch the rod to the pH paper.) Upon the addition of acid, a precipitate should form.

8. Collect the precipitate in a Büchner funnel by vacuum filtration (see figure in margin). Wash the precipitate with cold water (≈ 2 mL) and allow it to air dry.

9. Weigh the isolated carboxylic acid.

Isolation of the phenol

1. To the ether solution in the separatory funnel, add 5 mL of 1 M aqueous NaOH (sodium hydroxide solution).

2. Gently shake the separatory funnel for several minutes, venting it frequently.

3. Place the separatory funnel (uncapped) in a ring stand and allow the ether and aqueous layers to separate. Drain the lower aqueous layer into an Erlenmeyer flask.

4. Repeat the extraction with another 5 mL of 1 M aqueous NaOH. Combine the aqueous extracts. Save the ether layer in the separatory funnel for later use.

5. Heat the aqueous extract on a steam bath for 10 minutes to drive off any ether that is dissolved in the solution. The presence of excess ether will prevent precipitation of the phenol. After heating, cool the solution in an ice-water bath.

6. Once cooled, add 6 M HCl (≈ 2 mL) until the aqueous layer is acid. (Confirm acidity with pH paper.)

7. Collect the precipitate in a Büchner funnel by vacuum filtration. Wash it with cold water (≈ 2 mL) and allow it to air dry.

8. Weigh the isolated phenol.

Isolation of the alcohol

1. Add 10 mL of saturated sodium chloride solution to the ether in the separatory funnel and shake gently.

2. Allow the layers to separate and discard the lower sodium chloride layer.

3. Pour the ether layer into a beaker containing approximately 1 g of anhydrous Na_2SO_4 and allow to stand for about 15 minutes.

4. Decant the ether into a **tared** test tube and evaporate the ether by using a stream of air directed into the test tube.

5. When all of the ether has evaporated, weigh the remaining solid. *Note:* If oil forms instead of a solid, cool the sample on ice until it precipitates.

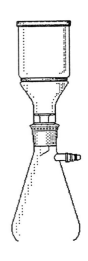

Büchner funnel (above) and side-arm vacuum flask (below).

Anhydrous sodium sulfate is used to remove any residual water from the ether.

Tare is the weight of the container. If the weight of the container is determined prior to adding a chemical, you can determine the weight of chemical by subtracting the tare from the total weight (container + chemical.)

PRE-LAB ASSIGNMENT

Fill in the following flowchart before coming to lab.

Mixture contains the following three compounds. Indicate in the appropriate boxes which compound is present at each stage of the separation.

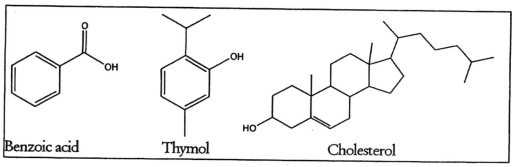

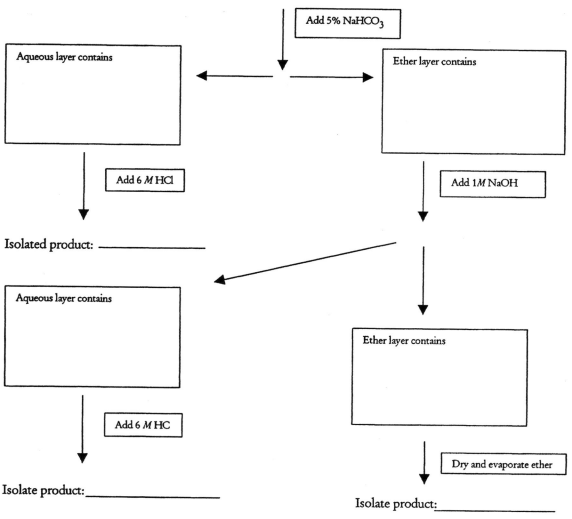

NAME_____

DATA AND RESULTS

1. Initial mass of the mixture: _____

2. Mass of each component after separation

   benzoic acid   _____

   thymol   _____

   cholesterol   _____

3. Calculate the percent recovery from the extraction, assuming the ratio of the mixture was 1:1:1 by weight.

$$\% \text{ recovery} = \frac{\text{mass recovered}}{\text{initial mass}}$$

Percent recovery benzoic acid: _____

Percent recovery thymol: _____

Percent recovery cholesterol: _____

4. One indication of purity if a sharp melting point. If you have the opportunity, take the melting points for the three isolated compounds. If the compounds are not yet dry, you may need to take the melting points after they have had a chance to dry. Report the melting points below.

QUESTIONS

1. What was the percent recovery of the initial mixture?

2. If the percentage was different from 100% (which in most cases it is), please comment on what might have caused the difference?

3. What would have happened to your separation if you had accidentally used the sodium hydroxide solution in the first separation instead of the sodium bicarbonate solution?

4. Why was the ether solution treated with anhydrous Na_2SO_4 before the ether was evaporated?

Experiment 23

The Synthesis of an Ester

In this experiment, you get to choose the reactants.

In this experiment, you will synthesize, and purify, an ester from an alcohol and carboxylic acid of your choice. Esters are interesting compounds and are major components of the odor and taste of fruits. Some esters and their odors are listed below.

n-propyl acetate	pear	isobutyl propanoate	rum
isopentyl acetate	banana	n-octyl acetate	orange
methyl salicylate	wintergreen	methyl anthranilate	grape
benzyl acetate	peach	isobutyl propanoate	floral
ethyl propanoate	fruity	propyl propanoate	pineapple
butyl benzoate	balsamic	methyl benzoate	prune
ethyl benzoate	fruit	geranyl acetate	rose
ethyl heptanoate	fruity	3-hexyl propanoate	Jolly Rancher apple

The equation below depicts the Fischer esterification reaction for the synthesis of an ester. A Fischer esterification is the acid-catalyzed synthesis of an ester from an alcohol and a carboxylic acid

$$\underset{\text{Acid}}{R-\overset{O}{\underset{\|}{C}}-OH} + \underset{\text{Alcohol}}{R'-OH} \underset{}{\overset{H_2SO_4}{\rightleftharpoons}} \underset{\text{Ester}}{R-\overset{O}{\underset{\|}{C}}-O-R'} + H_2O$$

To determine what alcohol and carboxylic acid you need in order to synthesize your ester, you have to remember how to name an ester. In an ester's name, the first word is the name of the group derived from the alcohol. The second word is the name of the acid, where "-ic" has been dropped and "oate" (or "ate") has been added. For example, methyl anthranilate is synthesized from methanol (methyl) and anthranilic acid (anthranilate).

Your instructor will give you a list of the alcohols and carboxylic acids that are available for use in the synthesis. Once you determine which ester you want to synthesize, you will need to devise the procedure for synthesizing approximately 20 mmole of the ester (assume a 100% yield). The procedure must be completed and checked before you can begin your work in laboratory.

Before writing your procedure, consider the following:

1. In this reaction, the acid and alcohol reactants are in equilibrium with the ester product (implied by the double arrow). In other words, once you reach equilibrium, there could potentially be a large amount of starting material remaining. This situation would result in a low yield of the ester. To resolve this problem, you can make use of Le Chatelier's principle. One of the consequences of Le Chatelier's principle is that we can drive the equilibrium to the right (to the products) by having one of the reactants in excess (in effect, putting *pressure* on the left side). Usually, a threefold molar excess of one of the reagents is sufficient to drive the equilibrium to the right in an esterification reaction.

    Le Chatelier's principle states that if a system at equilibrium is disturbed, the system will adjust to counteract the disturbance.

2. Note that sulfuric acid (H_2SO_4) is written over the equilibrium arrow. This notation means that sulfuric acid is used as a catalyst for this reaction; it will accelerate the rate at which the product is formed. Because a catalyst is not consumed during the course of a reaction, you need to use only a small amount of acid (3 or 4 drops) for it to be effective.

    Reflux, to heat so that the vapors formed condense and return to be heated again

3. Heating a reaction is another method used to increase the rate of the reaction. An organic reaction can be heated, but it is necessary to make sure none of the vapors escape from the reaction. To do this, we **reflux** the reaction mixture. Using a heat source, you reflux the solution below a water condenser, which will keep all the reagents and products from evaporating. There are many different types of reflux apparatus; they consist of a vessel for the reaction mixture attached to a condenser. An example of such an apparatus is shown in the margin.

4. You will be boiling an organic mixture. There are two general things you always do when boiling organic mixtures. You must insert a boiling stone before you begin heating the mixture. The boiling stone provides a surface upon which the vapor bubbles form and prevents a "bumping." The second thing is that you must never boil to dryness. Because you are refluxing the mixture, you should not lose any liquid. But as a cautionary note, watch the reaction closely to make sure the liquid is not evaporating.

5. Your product will need to be purified at the end of the reaction. Remember that you are adding one of the reagents in excess. You must consider how to remove the excess starting material. Most esters are not water soluble. If your excess reagent is water soluble, you can separate it from your product by "washing" the reaction mixture with water. Place the cooled reaction mixture in a separatory funnel with approximately 10 mL of ether. Add water to the funnel and shake gently. The excess water-soluble reagent will be in the water layer and separated from the ester. If neither reagent is water soluble you should add excess acid. The carboxylic acid can

Reflux apparatus

easily be converted into a water soluble carboxylate ion by washing with aqueous NaOH instead of with water.

6. Once you have developed a synthesis, create a reagent table for your reaction. For this you need find out the structures, physical properties, amounts used, etc., of all your reagents and products. This table will give you information on the reflux temperature (what components has the lowest boiling point), how to separate your product from excess reagents (what are the solubilities), and so on.

Fill in the pre-lab exercise and have your lab instructor check this procedure before you begin. Once this lab is complete, be sure to note the smells of your synthesized ester as well as the smells of the esters synthesized by your fellow classmates.

There are many Web sites where you might find useful information for this experiment. Two are listed below.

www.chemfinder.com This Web site allows you to put in a compound name and retrieve MW, boiling point, density (specific gravity), and other properties.

http://www.nysaes.cornell.edu/flavornet/index.html This Web site contains information on esters. Click on sidebar that says "Esters RI"; this will give you a list of compounds and their flavors/odors. If you click on a compound's name, you will see its structure. Not all of the compounds listed are esters, but this is an interesting site just the same. Just make sure that the compound you choose to synthesize is an ester.

NAME_____

PRE-LAB EXERCISE

1. What ester are you going to synthesize?

2. Give the name of the alcohol and the carboxylic acid needed to synthesize your ester.

    alcohol _____

    carboxylic acid _____

3. Write the reaction for the synthesis of your ester.

4. Find out the following information about your reagents and product. Blanks are provided for additional information.

Properties	Alcohol	Carboxylic acid	Ester
molecular formula			
molecular weight			
boiling point			
solubility			
density			

5. Calculate the amounts of reagents.

   a. The ratio of reagents to products is 1:1:1. So to synthesize 20 mmoles of products, you need 20 mmoles of each of the reagents. To calculate the grams needed, you multiply 20 mmoles (2.0×10^{-2} moles) by the molecular weight (MW).

   2.0×10^{-2} moles $\times$ MW (grams/mole) = grams of reagent (product)

   Grams of alcohol: _____

   Grams of carboxylic acid: _____

   b. Most of the alcohols and carboxylic acids you will use are liquids at room temperature. Rather than weigh out the liquid, you can measure out the volume of liquid that is equal to that weight. To do this you need to know the density of the liquid.

   $$\text{mL liquid} = \frac{\text{grams of liquid}}{\text{density of liquid (g/mL)}}$$

   If either of your reagents are liquids, calculate the milliliters of liquid required.

6. Now you need to make a decision: Which of the reagents are your going to add in excess? If you have a small alcohol that is water soluble, add that in threefold excess because it is easy to remove. If the alcohol is not water soluble, you will need to add the carboxylic acid in excess. You can remove excess carboxylic acid from your reaction mixture by converting it to a salt with a base. Once the carboxylic acid has been converted to a salt, it is water soluble and can be washed from the reaction mixture.

   What reagents are you going to add in excess? _____

   How much of this reagent do you need to add to have a threefold excess?

7. The last thing you need to determine is the approximate temperature for heating. This temperature will be determined empirically (by watching the reaction), but you need a starting point. You will want to start at the temperature of the lowest boiling component of your reaction mixture. You will start at this temperature and adjust the temperature such that the condensation ring (the place in the condenser where the vapor condenses to a liquid) if about half way up the condenser. What temperature are you going to start with?

Now you are ready to write the procedure. Write down the steps you are going to take to synthesize your ester. Start with the amounts you are going to mix together. Don't forget to add the catalyst and the boiling chips. Then you are going to heat the mixture for about 10 to 15 minutes— at what temperature? Lastly, write out how you are going to purify (remove excess reagent) from your product. Weigh your isolated product so that you can determine your yield. Be sure to have the lab instructor check over your procedure before you begin.

PROCEDURE

Equipment   Chemicals

NAME _____

DATA AND RESULTS

1. Mass of the ester synthesized: _____

2. Determine the percent yield for your synthesis

    a. What is the theoretical yield for your reaction?

    number of moles to be synthesized × MW of product = theoretical yield

    b. Determine the percent yield:

    $$\text{Percent yield} = \frac{\text{actual yield in grams}}{\text{theoretical yield in gram}} \times 100\%$$

3. Describe the smell of your ester. Based on this smell, do you think your synthesis was a success?

Synthesis of Aspirin

Americans consume approximately 80 billion aspirin tablets per year.

Willow bark tea is commonly used as a pain-killing agent.

Salicylic acid and its esters are found in several plants. The salicylic acid found in willow bark and leaves (*Salix* species) is the natural precursor for aspirin. The entire willow plant contains pain-killing constituents (analgesics), with the highest concentrations found in the inner bark.

The synthesis of aspirin from salicylic acid is an esterification reaction. Esters are formed from the acid-catalyzed reaction of an alcohol and a carboxylic acid. Salicylic acid (*o*-hydroxybenzoic acid) is a bifunctional compound. It is a phenol (an aromatic alcohol) and a carboxylic acid (R-COOH). Hence, it has the capacity of undergoing two different esterification reactions, acting as either the alcohol or the acid partner in the reaction. In the presence of acetic anhydride, acetylsalicylic acid (aspirin) is formed (reaction below). In this reaction, the acetic anhydride is reacting as if it were a carboxylic acid.

Salicylic acid   Acetic anhydride   Acetylsalicylic acid
                                              (aspirin)

In the reaction to prepare aspirin, the most likely impurity in the final product is salicylic acid itself. This impurity is generally the result of incomplete acetylation or hydrolysis of the product during the isolation steps. Salicylic acid, like most phenols, forms a highly colored complex with ferric chloride (Fe^{3+} ion). Aspirin, which has the hydroxyl group acetylated, will not give the color reaction. Thus, the presence of this impurity in the final product is easily detected.

is an acetyl group.

MATERIALS

Equipment

- ✓ 125-mL Erlenmeyer flask
- ✓ 400-mL beaker
- ✓ heat source
- ✓ thermometer
- ✓ dispensing pipet
- ✓ crystallizing dish
- ✓ glass rod
- ✓ 10-mL graduated cylinder
- ✓ 100-mL graduated cylinder
- ✓ Büchner funnel
- ✓ filter flask
- ✓ test tubes (3)

Chemicals

- ✓ salicylic acid
- ✓ acetic anhydride
- ✓ conc. sulfuric acid
- ✓ 4% phenol solution
- ✓ 2.5% ferric chloride solution

CAUTION!
Concentrated sulfuric acid is highly corrosive. You must handle it with great care.

PROCEDURE

Synthesis of aspirin

1. Using your 400-mL beaker set up a water bath and begin heating with a heat source. You want the water to reach a temperature of about 75°C.

2. While the water is heating, weigh 2.0 g (0.015 mole) of salicylic acid crystals and place them in a 125-mL Erlenmeyer flask. Add 5 mL (0.05 mole) of acetic anhydride, followed by 5 drops of concentrated sulfuric acid and gently swirl the flask until the salicylic acid dissolves.

3. Heat the flask gently in the water bath for 10 to 15 minutes.

4. Allow the flask to cool to room temperature, during which time the acetylsalicylic acid should begin to crystallize from the reaction mixture. If it does not, scratch the inside walls of the flask with a glass rod.

5. Cool the mixture slightly in an ice bath until complete crystallization has occurred. If crystallization has failed to occur, you may also use a seed crystal (obtain from instructor) to create a nucleus for crystal formation.

Scratches on the inside of the flask create a rough surface upon which crystallization can occur.

6. After crystallization is complete, add 50 mL of water and cool the mixture in an ice bath. Usually the product will appear as a solid mass when crystallization has become complete.

7. Collect the product by vacuum filtration, using a Büchner funnel (see figure). The filtrate can be used to rinse the Erlenmeyer flask repeatedly until all crystals have been collected.

8. Rinse the crystals several times with small portions of cold water. Continue drawing air through the crystals for air drying.

9. Weigh the crude product, which may contain some unreacted salicylic acid.

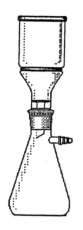

Büchner funnel (above) and side-arm flask (below).

Testing purity of product

1. To test the purity of your product set up three test tubes as follows:

   Test tube 1   5 mL water + 10 drops 4% phenol

   Test tube 2   5 mL water + a few crystals salicylic acid

   Test tube 3   5 mL water + a few crystals of product

2. Add 10 drops of 2.5% ferric chloride solution to each tube and note the color. Formation of an iron-phenol complex with Fe(III) gives a color ranging from red to violet, depending on the particular phenol. The presence of color in test tube 3 is an indication of an impurity.

3. Another method used to assay purity is to determine a melting point. Pure aspirin will have a sharp melting point of 135°–136°C. A broad and depressed (lower) melting is an indication of the presence of impurities.

Your instructor may ask you to save the product from this reaction to use in the next experiment, where you can examine the success of your synthesis of aspirin by using thin-layer chromatography (TLC).

NAME_____

RESULTS

Synthesis of aspirin

1. Record the masses of salicylic acid used and of product formed:

    Mass salicylic acid: _____

    Mass aspirin: _____

2. Calculate the theoretical yield of acetylsalicylic acid. The **theoretical yield** is the potential mass of product. It is based on the mass of reactants and the stoichiometry of the reaction. The stoichiometry of the reaction is 1 mole salicylic acid reacts with 1 mole acetic anhydride to form 1 mole aspirin.

    a. Calculate the moles of salicylic acid.

    $$\text{Moles salicylic acid} = \frac{\text{mass salicylic acid}}{\text{MW salicylic acid}}$$

*The **limiting reagent** is the reactant that is completely consumed in a reaction.*

    b. In this experiment, you used 0.05 moles of acetic anhydride. Which reactant is the limiting reagent, salicylic acid or acetic anhydride?

    c. To determine the theoretical yield, you need to multiply the molecular weight of aspirin by the moles of limiting reagent.

    Theoretical yield = (moles limiting reagent) × (MW acetylsalicylic acid)

3. Determine the percent yield of product.

$$\% \text{ yield} = \frac{\text{mass of product}}{\text{theoretical yield}} \times 100$$

Purity of product

1. Record the results of the ferric chloride test.

Test tube	Color
1	
2	
3	

2. On the basis of the results of the ferric chloride test, comment on the purity of your product.

3. If you determined the melting point of your product, record below. What does your melting point tell you about the purity of your product?

NAME_____

QUESTIONS

1. Why was sulfuric acid added to the reaction mixture?

2. Give some reasons why your yield was not 100%. If your yield was greater than 100%, give a potential cause.

3. As stated in the introduction, salicylic acid is a bifunctional compound. Draw and name the product of the esterification reaction of salicylic acid and methanol.

Experiment 25

Thin-Layer Chromatography of Analgesics

Many native peoples have used natural analgesics such as salicylic acid and its derivatives to relieve pain, reduce inflammation, and lower fevers.

In this experiment, you will use a separation technique called thin-layer chromatography (TLC). TLC is often used in organic chemistry to monitor the progress of a reaction or to identify the components of a mixture. In TLC, a small drop of the mixture to be separated is applied near one end of a flexible plate coated with silica gel. The end of the plate is then dipped into a developing solvent, called the **mobile phase**, which flows up the plate by capillary action. As the developing solvent flows up the plate, it can carry along the components of the mixture. The rate at which a particular component moves depends on whether it tends to dissolve in the developing solvent or to remain adsorbed (or stuck) on the surface of the silica gel, called the **stationary phase**. A component that moves rapidly is spending more time dissolved in the developing solvent, whereas a component that moves slowly is spending more time adsorbed to the silica.

The results of a chromatographic separation are expressed in terms of R_f values. An R_f is the *relative* distance that a sample component has moved, that is, relative to the distance moved by the developing solvent. As an example of R_f calculation, let's use the diagram of a typical chromatogram shown below.

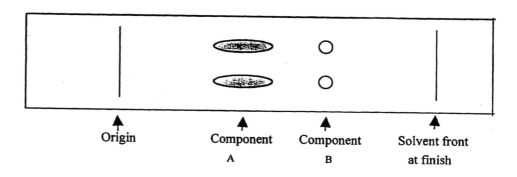

Distance is measured from the point where the sample is applied (the origin) to the middle of each component spot. It is also necessary to measure the distance from the origin to the location of the solvent front. In the above example, Component A traveled 4 cm and the developing solvent traveled 10 cm, the R_f of component A is calculated as follows.

$$R_f(A) = \frac{\text{distance A traveled}}{\text{distance solvent traveled}} = \frac{4 \text{ cm}}{10 \text{ cm}} = 0.4$$

R_f values determined under identical conditions are reasonably constant, so your calculated R_f values can be compared with another person's data or known values.

In this experiment, you will use TLC to identify the compounds found in analgesic tablets. Depending on the brand of analgesic, the active ingredient is most likely aspirin, acetaminophen, or ibuprofen. In some cases, caffeine is added as a stimulant. You will use known samples of acetaminophen, acetylsalicylic acid (aspirin), caffeine, and ibuprofen and compare them by TLC with the mixtures found in the analgesic tablets. The structures of these reference compounds are shown below.

An analgesic is a compound capable of relieving pain.

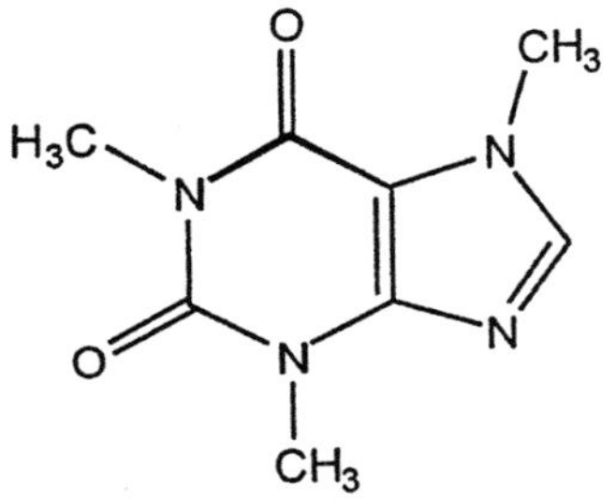

Caffeine

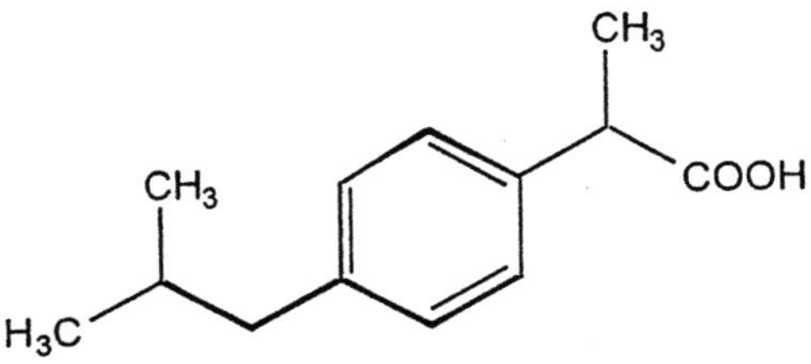

Ibuprofen

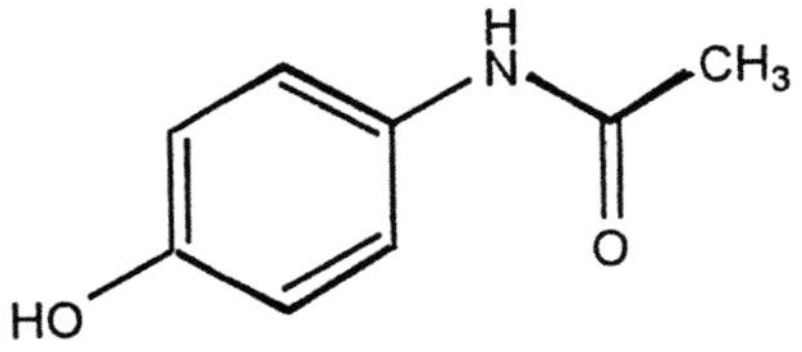

Acetaminophen

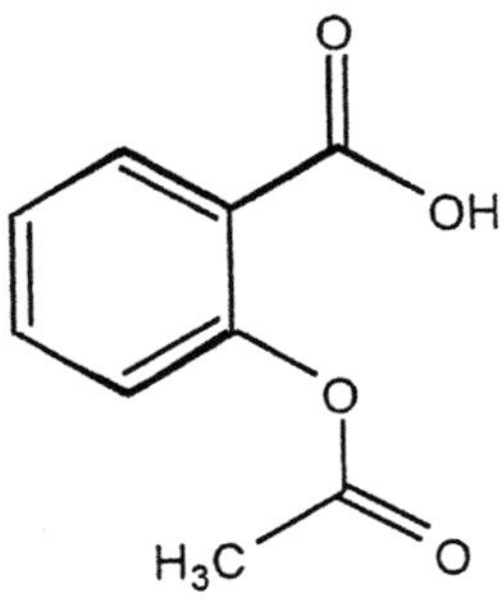

Aspirin

The method used to visualize the compounds will be illumination with ultraviolet (UV) light. The TLC plates are coated with a silica gel containing a fluorescent indicator. Under UV light, some of the spots will appear as dark areas on the plate, whereas others will fluoresce brightly. The differences in appearance under UV illumination, along with the distances the spots travel, will help to distinguish the substances from one another.

MATERIALS

Equipment

- ✓ 400-mL beaker
- ✓ test tubes
- ✓ stirring rod
- ✓ mortar and pestle
- ✓ water bath
- ✓ watchglass
- ✓ TLC plates
- ✓ UV lamp
- ✓ micro capillary tubes

Chemicals

- ✓ 5% solution acetaminophen
- ✓ 5% solution aspirin
- ✓ 5% solution caffeine
- ✓ 5% solution ibuprofen
- ✓ assortment of analgesics
- ✓ methylene chloride (CH_2Cl_2)
- ✓ ethanol (CH_3OH)
- ✓ 0.5% acetic acid in ethyl acetate
- ✓ ethyl acetate

PROCEDURE

1. Obtain a TLC plate, handle the plate carefully by the edges and do not bend. Using a lead pencil (not a pen), *lightly* draw a line across the plate (about 1 cm from the bottom. Using a centimeter ruler, lightly mark off six intervals equally spaced on the line. These are the points at which the samples will be spotted. You will spot two analgesics samples and each reference compound: acetaminophen, aspirin, caffeine, and ibuprofen.

2. Obtain half a tablet of each of the two analgesics to be analyzed. Crush the tablet by using a mortar and pestle; or you can crush it on a piece of filter paper with a spatula. Transfer each crushed half-tablet to a small, labeled test tube.

3. Using a graduated cylinder, mix together 10 mL of absolute ethanol and 10 mL of methylene chloride. Mix the solution well. Add 5 mL of this solvent to each of the test tubes and then heat each of them *gently* for a few minutes in a water bath. Not all the tablet will dissolve, because the analgesics usually contain an insoluble binder and/or buffering agents and coatings.

4. After heating the samples, allow them to settle and then spot the clear liquid extracts on the plate. To spot the plate, dip a capillary tube into the extract. Lightly touch the capillary tube to the plate. Allow just a small spot of liquid onto the plate.

5. Next, spot the four reference solutions—acetaminophen, aspirin, caffeine, and ibuprofen—on the plate.

6. When the plate has been spotted, it is ready to be placed in a development chamber. For a development chamber, you will use a 400-mL beaker lined with a piece of filter paper and covered with a watch glass. The purpose of the filter paper is to provide an atmosphere in the developing chamber that is saturated with solvent.

7. When the development chamber has been prepared, obtain a small amount of the development solvent (0.5% glacial acetic acid in ethyl acetate). Fill the chamber with the development solvent to a depth of about 1/4 inch. Be sure that the liner is saturated with the solvent. The solvent level must not be above the spots on the plate or the samples will dissolve off the plate into the reservoir instead of developing. Place the spotted plate in the chamber and allow the plate to develop.

8. When the solvent has risen to a level about 1 cm from the top of the plate, remove the plate from the chamber and, using a lead pencil, mark the position of the solvent front. Let the plate dry. When the plate is dry, observe it under a short-wavelength UV lamp. Lightly outline all the observed spots with a pencil.

RESULTS

1. Name of tested analgesics: 1 _____

    2 _____

2. Draw a diagram of the chromatogram.

```
┌─────────────────────────────────────────────────────┐
│                                                     │
│                                                     │
│                                                     │
│                                                     │
│                                                     │
│                                                     │
│                                                     │
│                                                     │
└─────────────────────────────────────────────────────┘
```

3. Record any observations about the appearance of the spots under UV illumination.

4. Report the R_f value for the reference compounds.

Compound	R_f value
acetaminophen	_____
aspirin	_____
caffeine	_____
ibuprofen	_____

5. Calculate the R_f values for the components of your tested analgesic tablets. Use the R_f values and observations to identify the components

                              R_f values   Identity of components

Tablet 1:

Tablet 2:

QUESTIONS

1. In examining your TLC of the analgesic tablets, did you see components other than the reference compounds? Explain.

2. In the introduction, we stated that TLC could be used to follow the progress of a reaction. Speculate on how this can be accomplished.

Extraction of Caffeine from Beverages

A look at the controversial compound, caffeine

Few compounds consumed by Americans is surrounded by as much controversy as caffeine. One article tells us that caffeine will cause anxiety, whereas another article tells us that the consumption of a cup of coffee before an exam will improve our grade. Approximately 50% of Americans drink coffee daily at a rate of over three cups per day. Caffeine is also found in soda pop, tea, chocolate, and pills that keep you awake. In addition to caffeine, tea and chocolate contain a related compound, theobromine. While the effects of theobromine are mild when compared to caffeine, it is believed that the effects of theobromine are enhanced by the presence of the caffeine in chocolate. The structures of caffeine and theobromine are shown below.

Caffeine   Theobromine

Caffeine is one of the oldest known stimulants. Caffeine stimulates both the central nervous system and skeletal muscles. This stimulation results in increased alertness, which is believed to be the reason for increased ability to concentrate, in the short term. The increased alertness also aids in the ability to stay awake when tired. The downside to this effect is that excessive intake of caffeine may result in restlessness, insomnia, and irritability. Heavy caffeine use can lead to both a tolerance to and a dependence on the consumption of caffeine. Anyone who has tried to kick the caffeine habit can tell you about the headache from caffeine withdrawal. The symptoms are real and in extreme cases can be so severe that the symptoms can include nausea and lethargy.

Check your caffeine consumption by examining the list of common caffeine-containing products below. How much caffeine do you consume in a day?

Product	Caffeine content
coffee (7 oz)	90–150 mg
espresso (2 oz)	100 mg
stay-awake pill	100 mg
instant coffee	60–80 mg
tea	30–70 mg
cola	30–45 mg
chocolate bar	30 mg
cold relief tablet	30 mg
chocolate milk (8 oz)	8 mg

In this experiment, you will extract the caffeine from some typical caffeine-containing products. The extraction will be a crude one, but it will give you an indication of the varying amounts of caffeine found in different products. You may also be asked to look at the purity of your isolate.

You can analyze the purity of your isolate caffeine by thin-layer chromatography (TLC). TLC is often used in organic chemistry to identify the components of a mixture. In TLC, a small drop of the mixture to be separated is applied near one end of a flexible plate coated with silica gel. The end of the plate is then dipped into a developing solvent, called the **mobile phase**, which flows up the plate by capillary action. As the developing solvent flows up the plate, it can carry along the components of the mixture. The rate at which a particular component moves depends on whether it tends to dissolve in the developing solvent or to remain adsorbed (or stuck) on the surface of the silica gel, called the **stationary phase**. A component that moves rapidly is spending more time dissolved in the developing solvent, whereas a component that moves slowly is spending more time adsorbed to the silica.

The caffeine in your isolate sample should move with the same speed as a reference sample of pure caffeine. If any additional components (impurities) are present in your isolate, you will see other spots present on the TLC plate.

MATERIALS

Equipment

- ✓ 250-mL Erlenmeyer flask
- ✓ 50-mL Erlenmeyer flask
- ✓ mortar and pestle
- ✓ 100-mL graduated cylinder
- ✓ spatula
- ✓ funnel
- ✓ test tube
- ✓ 250-mL beaker
- ✓ watchglass
- ✓ TLC plates
- ✓ UV lamp
- ✓ microcapillary tubes

Chemicals

- ✓ tea bags
- ✓ Surge, Dr. Pepper, or other caffeinated beverage
- ✓ Instant coffee
- ✓ No-Doz™ tablets
- ✓ methylene chloride
- ✓ Na_2CO_3
- ✓ anhydrous $MgSO_4$
- ✓ ethyl acetate

PROCEDURE

Part A: Isolation of caffeine

1. You will need 100 mL of beverage for the experiment. Choose one of the possible samples. For the coffee and tea, make up the equivalent of a strong cup of tea or coffee. For the soda pop, measure out 100 mL. Crush the No-Doz™ tablet, using the mortar and pestle, and dissolve the powder in 100 mL of water.

2. Add 2 g of sodium carbonate to your sample and swirl to dissolve.

3. Add 25 mL of methylene chloride (CH_2Cl_2) to the solution and gently stir for 5 to 10 minutes.

4. Decant off as much of the water layer (on top) as possible. The organic layer (methylene chloride plus caffeine) should remain in the Erlenmeyer flask.

5. Add anhydrous $MgSO_4$ to the methylene chloride solution. Stopper the flask and allow the solution to stand for 10 minutes. The magnesium sulfate will absorb any residual water in the solution.

6. Set up a filter funnel with folded filter paper. Filter the solution into a clean, tared 50-mL Erlenmeyer flask.

7. Evaporate the methylene chloride in the hood. The evaporation can be accomplished by using a hot plate set to a low heat setting. When the solution gets to about 1/16 inch, remove and let the rest evaporate. Another method for evaporation is to carefully blow a stream of air into the solution.

8. Reweigh the beaker and determine the amount of caffeine.

Part B: Analysis of the purity of the isolated caffeine

1. Obtain a TLC plate, handle the plate carefully by the edges and do not bend. Using a lead pencil (not a pen), *lightly* draw a line across the plate (about 1 cm from the bottom. Using a centimeter ruler, lightly mark off two intervals equally spaced on the line. These are the points at which the samples, your isolated caffeine and reference caffeine, will be spotted.

2. Transfer several crystals of your isolated caffeine to a small, labeled test tube.

3. Add 5 mL of methylene chloride, CH_2Cl_2, to the test tube and swirl gently to dissolve the crystals.

4. Spot the solution of your isolate on the plate. To spot the plate, dip a capillary tube into the solution. Lightly touch the capillary tube to the plate. Allow just a small spot of liquid onto the plate.

5. Next, spot the caffeine reference solution on the plate.

6. When the plate has been spotted, it is ready to be placed in a development chamber. For a development chamber, you will use a 250-mL beaker lined with a piece of filter paper and covered with a watch glass. The purpose of the filter paper is to provide an atmosphere in the developing chamber that is saturated with solvent.

7. When the development chamber has been prepared, obtain a small amount of the development solvent (ethyl acetate). Fill the chamber with the development solvent to a depth of about 1/4 inch. Be sure that the liner is saturated with the solvent. The solvent level must not be above the spots on the plate or the samples will dissolve off the plate into the reservoir instead of developing. Place the spotted plate in the chamber and allow the plate to develop.

8. When the solvent has risen to a level about 1 cm from the top of the plate, remove the plate from the chamber and, using a lead pencil, mark the position of the solvent front. Let the plate dry.

9. When the plate is dry, observe it under a short-wavelength UV lamp. Lightly outline all the observed spots with a pencil. Also observe under ordinary light and circle any spots.

NAME _____

DATA AND RESULTS

Part A: Isolation of caffeine

Sample chosen (beverage, brand): _____

Amount of sample used: _____

1. Determine the amount of caffeine isolate.

    Weight of flask with caffeine _____

    Weight of empty flask _____

    Weight of isolated caffeine _____

2. Describe the appearance of the isolated caffeine.

Part B: Analysis of the purity of the isolated caffeine

1. Draw a diagram of the chromatogram.

2. Record any observations about the appearance of the spots under ordinary light and UV illumination.

QUESTIONS

1. How did you results compare with the expected caffeine amounts listed in the table? Can you account for any difference?

2. From the appearance of your isolated caffeine, do you believe that your sample was pure caffeine? If not, what do you think are the possible contaminants?

3. Did the TLC provide you with any additional information about the purity of your isolated caffeine? Explain.

Experiment 27

Biochemistry Laboratory Techniques

A look at skills required for working in a biochemistry lab

One of the most important and time-consuming aspects of laboratory research is making solutions. This activity is particularly crucial in biochemical experiments where pH, ionic strength, and substrate concentration can be significant factors. No matter how carefully you plan the experiment, if the solutions are improperly prepared, the experiment may fail. Preparing solutions is a basic and essential skill. In Seattle, biotechnology companies state the number one basic skill expected of new employees in research positions is the ability to prepare solutions, buffers, and media. The preparation of a solution can be included as part of the job interview.

In the laboratory, you are often required to make a variety of solutions and dilutions. One interesting aspect of the solutions used in biochemistry labs is that procedures often mix concentration units. For example, phosphate buffered saline, PBS, is a reagent often used in research with platelets; it is 10 mM phosphate buffer, pH 8.0, containing 0.9% (w/v) NaCl. A solution of this type requires that you be facile in making both molar and percentage solutions.

> Molar solutions are given in molarity (M). Molarity is moles per liter of solution.

Giving concentration units in percentages is common in both biochemical and medical research. Percentage solutions are given as weight per volume (w/v) for solids and volume per volume (v/v) for liquids. A 10% aqueous NaCl solution has 10 g of NaCl in a final volume of 100 mL. A 10% aqueous ethanol (CH_3CH_2OH) solution is 10 mL of ethanol diluted to 100 mL with water.

In this laboratory exercise, you will make a series of solutions and test your accuracy by taking the pH of the solution. In addition, you will create a standard curve and use it to determine the concentration of an unknown solution. Another thing you will find in a biochemistry laboratory is the use of small units. In chemistry experiments, you have mostly dealt with volumes in liters and milliliters; but in biochemistry, you often use the unit microliters (μL) or concentrations that are in the nanomolar (n*M*) range. You must be able to convert between μL and mL easily.

milli (m) = 10⁻³
micro (μ) = 10⁻⁶
nano (n) = 10⁻⁹

To get you off on the right foot for the biochemistry portion of the course, you will perform several of these necessary tasks. This exercise will give you a chance to practice these skills before having to do them in the context of an experiment. In the lab practical, you will be required to make solutions, take measurements, graph, and do several conversions from one unit to another. To give you a head start, perform as many of the calculations as possible before coming to lab.

For example, Task 1 is the preparation of 100 mL of each of the primary standards used for calibration of pH meters. The three standards are

pH (25°C) = 4.01; 0.05 *M* potassium hydrogen phthalate

pH (25°C) = 6.86; 0.025 *M* potassium dihydrogen phosphate/0.025 *M* disodium hydrogen phosphate

pH (25°C) = 9.18; 0.01 *M* sodium tetraborate

Before coming to lab, you should make all the calculations required for preparing these solutions. Molecular weights for these compounds can be found in either a CRC manual (*Handbook of Chemistry and Physics* or *Handbook of Biochemistry*), a *Merck Index*, an Aldrich catalog, or a Sigma catalog. Before making the solutions in the lab, be sure to read the label! Some of the salts might be hydrated, and you may need to adjust your calculations to compensate for the hydration. To help you with this task we will outline the calculations involved for the preparation of the pH 9.18 standard.

Sodium tetraborate, $Na_2B_4O_7$, can be found in either the anhydrous or the hydrated form. Anhydrous sodium tetraborate has a formula weight of 201.2 g/mol. The hydrated form is a decahydrate; that is, each molecule of sodium tetraborate has ten molecules of water associated with it. Therefore, when determining the formula weight, you need to include the weight of the associated water. Sodium tetraborate decahydrate (written $Na_2B_4O_7 \cdot 10\, H_2O$) has a formula weight of 381.4 g/mol.

Anhydrous means "without water."

How much anhydrous sodium tetraborate is required to make 100 mL (0.1 liters) of 0.01 *M* solution?

$$0.01\ M \text{ solution} = \frac{0.01\ \text{mole}}{1\ \text{liter}} = \frac{0.001\ \text{mole}}{0.1\ \text{liter}}$$

$$0.001\ \text{moles anhydrous } Na_2B_4O_7 = 0.001\ \text{mole} \times \frac{201.2\ \text{g}}{\text{mole}} = 0.2012\ \text{g}$$

$$0.01\ \text{moles } Na_2B_4O_7 \cdot 10\, H_2O = 0.001\ \text{mole} \times \frac{381.4\ \text{grams}}{\text{mole}} = 0.3814\ \text{g}$$

The last two tasks (Task 4 and 5) in the lab practical required you to graph data. You might want to review the Appendix, which gives you information on graphing. Task 4 can be completed before coming to lab.

In Task 5 you will make a serial dilution and then measure the absorbance of the solutions. A **serial dilution** is a process of making dilutions in series. That is, use solution A to make a more dilute solution B, then use solution B to make an even more dilute solution C, and so on. This is in contrast to starting from scratch with the solute for each solution. There is a benefit in making serial dilutions in that you only need measure the solute once and from thenceforth you use the initial concentration to make all subsequent concentrations. This process is also the risk in making serial dilutions, because if you make a mistake in making the initial solution then all subsequent solutions are incorrect.

$C_1V_1 = C_2V_2$
C is concentration
V is volume

Before coming to lab, determine the concentrations of the solutions from the information given in Task 5. Let's go through one calculation so that you can see how to determine the concentrations. You should use the formula: $C_1V_1 = C_2V_2$ to determine the concentrations of the serial dilution. Test tube 1 contains a solution that has a concentration of 7.5×10^{-5} M. You will take 3 mL (V_1) of 7.5×10^{-5} M (C_1) solution and put it into test tube 2. To this 3 mL you add 1.5 mL of water for a final volume of 4.5 mL (V_2). What is the final concentration (C_2)?

$$C_1V_1 = C_2V_2 \qquad (7.5 \times 10^{-5} M)(3 \text{ mL}) = C_2 (4.5 \text{ mL})$$

$$22.5 \times 10^{-5} M \cdot mL = C_2 (4.5 \text{ mL})$$

$$C_2 = 22.5 \times 10^{-5} M \cdot mL / 4.5 \text{ mL} = 5 \times 10^{-5} M$$

Therefore the concentration of the solution in test tube 2 is 5×10^{-5} M. Perform similar calculations to determine the concentrations in test tubes 3 and 4.

MATERIALS

Equipment

- ✓ small test tubes
- ✓ large test tubes
- ✓ test-tube rack
- ✓ 100-mL volumetric flask
- ✓ pH meter
- ✓ 10-mL graduated cylinder
- ✓ spectrophotometer

Chemicals

- ✓ sodium tetraborate
- ✓ potassium dihydrogen phosphate
- ✓ disodium hydrogen phosphate
- ✓ potassium hydrogen phthalate
- ✓ 7.5×10^{-5} M ferroin

LAB PRACTICAL

Task 1

Each student will prepare 100 mL of each of the primary standards for calibration of pH meters. The three standards are

pH (25°C) = 4.01; 0.05 M potassium hydrogen phthalate

pH (25°C) = 6.86; 0.025 M potassium dihydrogen phosphate/
0.025 M disodium hydrogen phosphate

pH (25°C) = 9.18; 0.01 M sodium tetraborate

Use a pH meter to check the accuracy of your solutions. Report the pH of each solution below.

Solution	pH of your solution
0.05 M potassium hydrogen phthalate	_____
0.025 M potassium dihydrogen phosphate/ 0.025 M disodium hydrogen phosphate	_____
0.01 M sodium tetraborate	_____

Task 2

> A **chelate** is an organic compound that holds a metal ion by coordinate bonds.

Sample loading buffer is used in electrophoresis of DNA; the glycerol is a viscous liquid that makes the sample denser than water; the EDTA chelates divalent metal ions, which might cause interference; and the tracking dyes are there so that you can monitor the progress of the electrophoresis. *Describe* (don't actually make it) how you would make 10 mL of sample loading buffer that contains

30% (v/v) glycerol
0.25% (w/v) xylene cyanol FF
50 mM EDTA (ethylenediaminetetraacetic acid)
0.25% (w/v) bromophenol blue

Task 3. Conversions

1. Determine the amount of agarose (a solid) needed to make 50 mL of 1% agarose (w/v) solution _____ grams

2. Convert the following amounts to µL:

acrylamide/bis stock solution	2.5 mL	_____
10% SDS	0.1 mL	_____
distilled water	4.8 mL	_____
10% APS	0.1 mL	_____
TEMED	0.01 mL	_____

Task 4. Determination of the molecular weight (MW) of an unknown protein by gel electrophoresis

There is an inverse relationship between the migration distance of a protein in gel electrophoresis and the log of the molecular weight of that protein. Convert the following MWs to log values and plot on regular graph paper (log MW versus distance traveled). Use this standard graph to determine the MW of the unknown protein. Attach the graph before you hand in your lab practical.

*In biochemistry, the more archaic term— **molecular weight**— is used instead of molecular mass.*

Standard protein	MW (daltons)	Distance traveled (cm)
trypsin inhibitor	20,100	3.6
carbonic anyhydrase	29,000	3.1
alcohol dehydrogenase	39,800	2.7
serum albumin	66,000	2.1
unknown protein	??	2.4

MW of unknown protein _____

Task 5: Serial dilution and standard curve

To determine the concentration of an unknown chemical or protein, you often must prepare a standard curve. This task is done by making serial dilutions of known concentrations and determining the absorbance of each concentration. From this information, you can construct a graph that you can use to determine the concentration of a sample. The instructions for doing a serial dilution of ferroin solution are given below.

NAME_____

Determine the concentration of each sample from the information given. After doing the serial dilution, you will then determine the absorbance of each solution. The intensity of the color is proportional to the concentration of ferroin. Construct a graph of concentration ferroin versus absorbance.

1. Turn on the spectrophotometer and allow it to warm up for 10 minutes.

2. Label 4 large test tubes, 1 through 4.

3. Add 1.5 mL of water to test tubes 2 and 4 and 2.25 mL of water to test tube 3

4. Add 6.0 mL of ferroin standard (7.5×10^{-5} M ferroin) to test tube 1.

CAUTION!
Be sure to rinse your graduated cylinder between each dilution.

5. Using your graduated cylinder, add 3 mL of the solution from test tube 1 to test tube 2 and mix the contents.

6. Add 2.25 mL of the solution from test tube 2 to test tube 3 and mix the contents.

7. Add 1.5 mL of the solution from test tube 3 to test tube 4 and mix the contents.

8. Place 3 mL of each solution into four small test tubes. Place 3 mL of water into another small test tube to use as a reagent blank. The reagent blank is used to zero the spectrophotometer. Follow the directions for the particular spectrophotometer used in your laboratory.

9. Read and record the absorbance of all tubes at 508 nm.

10. Plot concentration of ferroin versus absorbance on the grid on the next page.

Test tube	Ferroin concentration	Absorbance
1		
2		
3		
4		

Plot absorbance versus concentration to create a standard curve for ferrion.

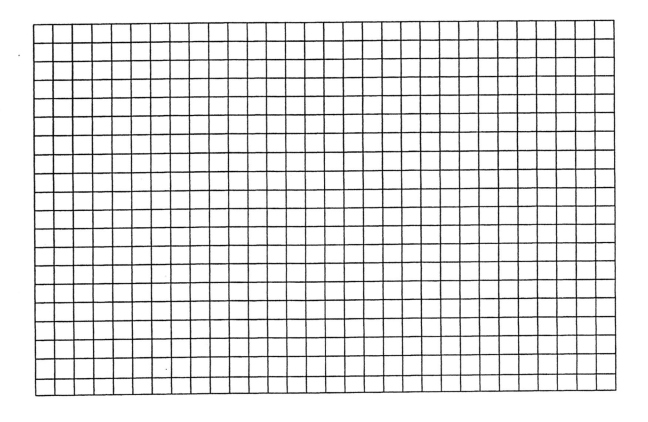

Separation of Glucose from Starch

One of the first steps in digestion is the breakdown of starch by an enzyme in saliva.

Monosaccharides and Disaccharides

Carbohydrates, also known as sugars or saccharides, are one of the most important classes of compounds dealt with in organic chemistry. These compounds function as sources of energy in living organisms and as structural material for cells. Carbohydrates are complex molecules containing both alcohol and carbonyl functional groups. One of the main sources of carbohydrates in our diet is starch from grains and cereals

Carbohydrates, such as starch, can also be viewed as polymers of various **monosaccharides**. A monosaccharide is the simplest type of carbohydrate; it cannot be further hydrolyzed into smaller units. One such example of a monosaccharide is glucose, which is shown below.

D-(+) Glucose (hemiacetal form)   D-(+) Glucose (aldehyde form)

Because glucose has both an alcohol and a carbonyl functional group, it can form a cyclic hemiacetal, as shown in the reaction above. The hemiacetal form is in equilibrium with the open chain form. Because both of these forms exist, one would expect that the chemistry of these sugars would be similar to that of an aldehyde. In the case of glucose, its aldehyde form can be easily oxidized. Carbohydrates that can be oxidized are known as **reducing sugars**.

A **disaccharide** is a molecule composed of two monosaccharide units. One such compound is sucrose (structure shown below), also known as table sugar. Sucrose is composed of a fructose unit and a glucose unit. Note that the fructose and glucose units are joined by an acetal link involving both their carbonyl (anomeric) carbons, which means that the aldehyde and ketone forms of these sugars are unavailable. Because sucrose is not in equilibrium with any aldehyde form, it is considered to be a **nonreducing sugar**.

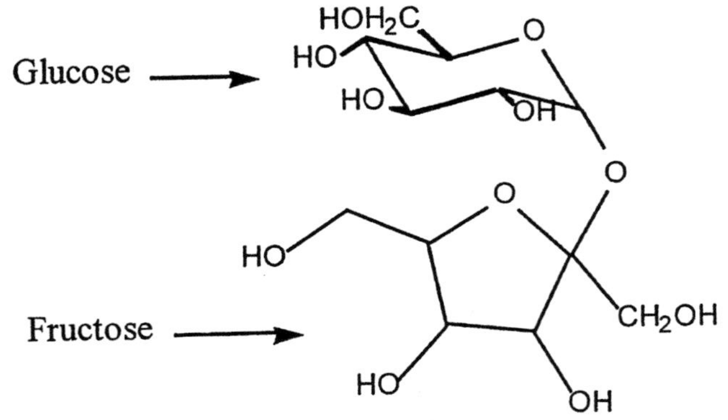

Glucose ⟶

Fructose ⟶

Sucrose

Two other common disaccharides are maltose (composed of two glucose units) and lactose (which contains glucose and galactose).

Longer chains (polymers called **polysaccharides**) of saccharides exist. The monomer glucose is used to form two different polymers, starch and cellulose. The difference between these two polymers is how the polymeric linkage is formed. We, as humans, can utilize starch as a source of glucose because we can hydrolyze the linkages between the glucose units in starch (α linkages). In contrast, we cannot use cellulose as a source of glucose because we do not contain the enzyme (β-glycosidase) to hydrolyze the β linkages found in cellulose.

Glucose polymer with α (1-4) glycosidic linkages

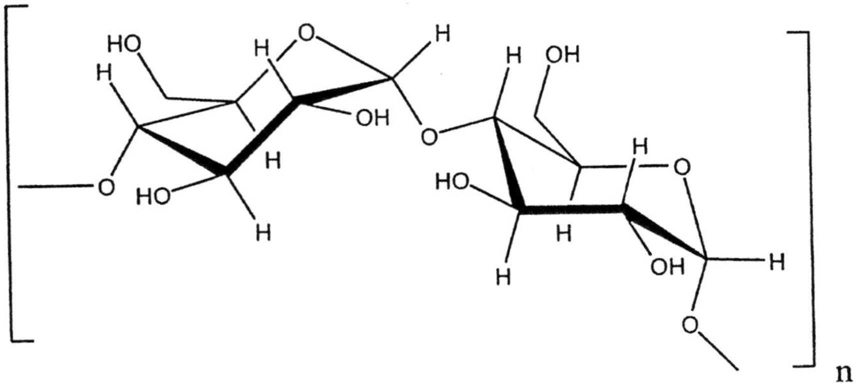

The α and β indicates the configuration at the anomeric carbon. The α-anomer has a trans relationship between the hydroxy on the anomeric carbon and the –CH$_2$OH at C5. In the β-anomer, the relationship is cis.

Glucose polymer with β(1-4) glycosidic linkages

[Chemical structure of glucose polymer]

In the first part of this experiment, you will use the process of gel filtration to separate glucose from starch in a 50:50 mixture. Separation by gel filtration depends on the different abilities of sample molecules to enter pores of gel. Molecules larger than the largest pores of the gel cannot enter the pores (they are excluded) and therefore pass rapidly down the column between the beads. The path of smaller molecules through the gel bed is much longer than that of the large molecules because they move in and out of the pores of the gel (they are included). Thus, the ability of the small molecules to penetrate the pores in the gel results in a retardation of their migration down the column.

Stages of a Separation of Two Molecules of Different Size on a Gel Filtration Column

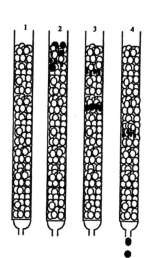

1. Hydrated beads (large open circles) are packed into a column to form the gel bed.

2. A sample containing a mixture of a large molecule (large black dots) and a small molecule (small black dots) is added to the top of the gel bed.

3. Buffer is applied to the top of the column and the molecules enter the gel bed. The smaller molecules penetrate the gel beads and their movement down the column is retarded. The larger molecules are excluded from the gel beads and move rapidly down the column.

4. The two molecules are now separated and can be collected in the column eluate.

In this experiment, the two molecules that are being separated are starch (large molecule) and glucose (smaller molecule). The fractions collected from the column will be tested for the presence of starch and glucose. The presence of starch can be detected by adding iodine to the solution. In the presence of starch, iodine changes to a blue-black color. Benedict's test will be used to test for the presence of glucose. In the Benedict's test, copper (II) is reduced to copper (I) by reducing sugars, a reaction resulting in a change in the color of the solution from pale blue to dark green to brick red color.

In the second part of the experiment, you will verify that humans can cleave (hydrolyze) starch. You will use the enzyme amylase to hydrolyze starch into maltose, a dimer of glucose. You will incubate an aliquot of starch with amylase and then use gel filtration to separate any residual starch from the maltose. Because maltose is a reducing sugar and will give a positive Benedict's test, by using this test you can determine whether any of the starch has been hydrolyzed.

MATERIALS

Equipment

- ✓ small test tubes (20)
- ✓ 10-mL graduated cylinder
- ✓ 250-mL beaker
- ✓ PD-10 gel filtration column
- ✓ hot plate
- ✓ thermometer
- ✓ test-tube rack

Chemicals

- ✓ 0.1% aqueous glucose solution
- ✓ 0.1% aqueous starch solution
- ✓ 0.5% aqueous starch solution
- ✓ 0.9% aqueous NaCl
- ✓ Benedict's reagent
- ✓ iodine solution
- ✓ aqueous amylase solution (crude, 1 mg/mL)

PROCEDURE

Part A: Separation of starch from glucose

1. Mount the PD-10 gel column vertically, using a ring stand and a clamp.

2. Number a set of test tubes from 1 to 10. Put 1 mL of water into one unlabeled test tube and mark the height. You will this tube as a guide as to how much eluant (liquid) to collect into each test tube. Place test tubes in test-tube rack.

3. Remove the top from the column. Place a small beaker under the column. Remove the stopper from the column tip and allow the eluant to drip out of the column until the flow stops.

4. Add 1 to 2 mL the 0.9% aqueous NaCl to the top of the column to flush the column. Leave column open until flow stops.

5. In a separate test tube mix 0.5 mL of the 0.1% glucose solution with 0.5 mL of the 0.1% starch solution.

6. Add the 1.0 mL starch:glucose mixture to the top of the gel column and continue to collect eluant in the beaker. Discard all eluant washings.

7. When the eluant flow has stopped, cap the outlet spout.

8. Add 0.9% aqueous NaCl solution to fill the top of the column and lower the gel column so that the spout is slightly above test tube 1.

9. Remove the outlet cap and collect 1 mL of eluant into each test tube 1 through 10. The column may need to be refilled with 0.9% aqueous NaCl during the fraction collection.

10. Cap the outlet spout after eluant in test tube 10 has been collected. Make sure the space above the gel column is approximately half full of NaCl solution before continuing.

Test for presence of starch

1. Number a second set of test tubes 1A to 10A. Add 1 drop of iodine to each test tube.

2. Remove a few drops of eluant collected in test tubes 1 to 10 into numerically corresponding test tubes 1A to 10A. Only a few drops of solution are needed for the starch test. The original test tubes will be tested for the presence of glucose.

3. The presence of starch is indicated by a change from the original brown iodine color to a blue-black color. Hold the tube against a white background to help you observe the color change.

Test for presence of glucose

Time saving tip: Start the hydrolysis of starch while doing the test for glucose.

1. For the glucose test: Place approximately 1 mL of Benedict's solution into each of the original test tubes, 1 to 10.

2. Gently heat these test tubes in a warm water bath for 10 minutes. A reducing sugar will produce a red, green, or yellow precipitate.

3. Place waste from the Benedict's tests in an appropriate waste container. All other waste can be disposed of down the drain.

4. Clean all 20 test tubes to be used in the next procedure.

Part B: Hydrolysis of starch with amylase

1. Place 1 mL of 0.5% aqueous starch solution into a test tube and add approximately an equal volume of amylase solution. Incubate the mixture in a warm water bath (50°C) for 10 to 15 minutes.

2. Remove the test tube from the water bath and allow it to cool to room temperature.

3. Remove the end cap from the column and flush the column with 10 mL of salt solution. This will wash out any residual material.

4. Add the 1.0 mL of hydrolyzed starch to the top of the gel column and continue to collect eluant in the beaker. Discard all eluant washings.

5. Repeat Steps 7 to 10 as described in the separation of starch from glucose, using the hydrolyzed starch sample.

6. Perform the iodine and Benedict's tests for the presence of starch and maltose on the eluted samples.

DATA

In the table below indicate a positive test with a (+) and a negative test with a (–).

Part A: Separation of starch and glucose

Test tube	Iodine test	Benedict's test
1		
2		
3		
4		
5		
6		
7		
8		
9		
10		

Part B: Separation of starch and glucose

Test tube	Iodine test	Benedict's test
1		
2		
3		
4		
5		
6		
7		
8		
9		
10		

QUESTIONS

1. Comment on the quality of the separation of glucose from starch? For example, was there any overlap between the two compounds?

2. From your results, can you determine whether amylase cleaved the starch?

3. Was there any residual starch in your hydrolyzed sample?

4. What results would you expect if you used cellulose in the hydrolysis experiment instead of starch?

Hydrolysis of Sucrose

You will compare the hydrolysis of sucrose by acid and by enzyme.

Carbohydrates, also known as sugars or saccharides, are one of the most important classes of compounds encountered in biochemistry. These compounds function as a source of energy for living organisms as well as providing structural material for cells. Carbohydrates are complex molecules containing both alcohol and carbonyl functional groups.

Carbohydrates can exist as single sugar units called **monosaccharides** or as polymers called **polysaccharides**. Two examples of monosaccharides are glucose and fructose, shown below.

D-(+) Glucose (hemiacetal form)   D-(+) Glucose (aldehyde form)

D-(−) Fructose (hemiketal form)   D-(−) Fructose (ketone form)

Because glucose and fructose have both alcohol and carbonyl functionalities they are able to form cyclic hemiacetals and hemiketals. In solution, the hemiacetal and hemiketal forms are in equilibrium with the open chain forms. Because both forms exist, the chemistry of these sugars is similar to that of a ketone in the case of fructose and an aldehyde in the case of glucose. The aldehyde form of glucose is easily oxidized. Carbohydrates that can be oxidized are known as **reducing sugars**.

In a redox reaction, the compound being oxidized is the **reducing agent**.

A **disaccharide** is a molecule composed of two monosaccharides. One such compound is sucrose, what you know as table sugar. Sucrose is composed of a fructose unit and a glucose unit. The fructose and glucose units are joined by an acetal linkage between both the carbonyl carbons. Unlike a hemiacetal or hemiketal, an acetal is stable in solution and the ring does not open. Because the ring does not open, the aldehyde and ketone forms are unavailable. Because sucrose is not in equilibrium with an aldehyde form, it is considered a **nonreducing sugar**.

Sucrose

Contrast sucrose with lactose, which is a disaccharide composed of a galactose and glucose unit. In lactose, a hemiacetal bond is formed with an alcohol oxygen of glucose, so the aldehyde functionality of the glucose unit is available and can be oxidized. Hence, lactose is a reducing sugar.

Galactose unit   Glucose unit   Galactose unit   Glucose unit

Lactose with cyclic glucose   Lactose with glucose open ring

As stated earlier, sugars are used by the body as a source of energy. To get energy from table sugar (sucrose), your digestive system must first break it into the monosaccharide units glucose and fructose. In the laboratory, this hydrolysis of the acetal linkage is normally done by heating sucrose in a strong acid solution, for example, 3 M HCl. However, the human body needs to maintain a neutral pH and cannot tolerate such acidic conditions. Fortunately, living organisms can hydrolyze carbohydrates to gain glucose by using enzymes. Enzymes are protein catalysts that accelerate reactions, such as the hydrolysis of the acetal linkage. One enzyme that can hydrolyze sucrose to glucose and fructose at pH 7 is invertase

The hydrolysis of sucrose

In this experiment, you will explore the hydrolysis of sucrose and lactose. You will first use Benedict's reagent to test sucrose, glucose, and lactose to see whether they are reducing sugars. In the Benedict's test, copper (II) is reduced to copper (I) by reducing sugars, a reaction that changed the color of the solution from light blue to colors ranging from dark green to brick red.

You will hydrolyze sucrose into fructose and glucose by the two different methods discussed above. One solution of sucrose will be heated in 3 M HCl and the other solution will be incubated with the enzyme invertase. To see whether your hydrolyses were successful, you will test for the presence of reducing sugars (inferring the presence of glucose and fructose) after the reactions has been completed. If sucrose is hydrolyzed, you will see a positive Benedict's test.

MATERIALS

Equipment

- ✓ small test tubes (6)
- ✓ test-tube rack
- ✓ 10-mL graduated cylinder
- ✓ 250-mL beaker
- ✓ thermometer

Chemicals

- ✓ sucrose
- ✓ lactose
- ✓ glucose
- ✓ invertase
- ✓ Benedict's reagent
- ✓ 3 M HCl

PROCEDURE

1. Add 0.3 g of sucrose to three different test tubes, and label the tubes 1 through 3.

2. To test tube 1, add 5 mL of distilled water (this will be a blank). To test tube 2, add 5 mL of 3 M HCl. To test tube 3, add 5 mL of distilled water and 2 mg of invertase.

3. Place the three test tubes in a water bath and heat at 37°C for 50 minutes. Shake tube 3 periodically.

4. While test tubes 1 to 3 are incubating, prepare three additional test tubes labeled 4 to 6. To a test tube (4), add 0.3 g of sucrose to 5 mL of distilled water. In another test tube (5), add 0.3 g of lactose to 5 mL of distilled water. In the last test tube (6), add 0.3 g of glucose to 5 mL of distilled water.

5. When the incubation of test tubes 1 to 3 is completed, perform the Benedict's test on all six samples. For the Benedict's test add 5 mL of Benedict's reagent to the solutions in each test tube (1 to 6)

6. Heat these mixtures on a hot water bath for 10 minutes. A reducing sugar will produce a red, green, or yellow precipitate; this is considered a positive test.

NAME _____

RESULTS

Test tube (sample)	Result (color)	Positive/negative
1 (Blank)		
2 (sucrose + HCl)		
3 (sucrose + invertase)		
4 (sucrose)		
5 (lactose)		
6 (glucose)		

Did you see hydrolysis of sucrose under both conditions? Do you see any difference between the two samples?

QUESTIONS

1. Honey is also known as invert sugar. In honey, the sucrose has been hydrolyzed into fructose and glucose. Is honey a reducing or nonreducing sugar? Explain.

2. Starch is a polysaccharide consisting of hundreds of units of glucose. Although glucose is a reducing sugar, starch is not. Explain this phenomena.

Synthesis of a Soap

Historically, soap was made from beef tallow (fat) and wood ashes.

Fats and oils are called triacylglycerols or triglycerides.

In this experiment, you will prepare soap from animal fat (lard). Animal fats and vegetable oils are esters of long chain carboxylic (fatty) acids and the alcohol, glycerol. Glycerol contains three alcohol functional groups and thus can react with three carboxylic acids. Soaps can be formed from animal fats and vegetable oils by the hydrolysis under alkaline conditions, saponification (reaction shown below).

$$
\begin{array}{c}
R_1\overset{O}{\underset{\|}{C}}-O-CH_2 \\
R_2\overset{O}{\underset{\|}{C}}-O-CH \\
R_3\overset{O}{\underset{\|}{C}}-O-CH_2
\end{array}
\xrightarrow[\text{or hydrolysis}]{\text{NaOH saponification}}
\begin{array}{c}
R_1COO^-Na^+ \\
R_2COO^-Na^+ \\
R_3COO^-Na^+
\end{array}
+
\begin{array}{c}
HO-CH_2 \\
HO-CH_2 \\
HO-CH_2
\end{array}
$$

Triglycerides (fat or oil) → Carboxylic acid salts (soap) + Glycerol

The fatty acids in a triglyceride are rarely of a single type in any given fat or oil. In fact, a single triglyceride molecule may contain three different acid residues (R_1COOH, R_2COOH, R_3COOH). Each fat or oil, however, has a characteristic distribution of the various types of acids possible. For example, beef tallow generally contains 50% oleic acid with the majority of the remaining acids being myristic (C_{14}), palmitic (C_{16}), and stearic (C_{18}) unsaturated acids.

Tallow is the principal fatty material used in making soap. Soapmakers usually blend tallow with coconut oil and this mixture is saponified. The resulting soap contains mainly the salts of palmitic, stearic, myristic, and oleic acids from the tallow, and the salts of lauric and myristic acids from the coconut oil. The coconut oil is added to produce a softer, more soluble soap.

Tallow $\quad CH_3(CH_2)_{14}COOH \quad CH_3(CH_2)_{16}COOH$

Palmitic acid $\qquad$ Stearic acid

$$CH_3(CH_2)_7CH=CH(CH_2)_7COOH$$

Oleic acid

Coconut oil $\quad CH_3(CH_2)_{10}COOH \quad CH_3(CH_2)_{12}COOH$

Lauric acid $\qquad$ Myristic acid

Depending on the source of the fat and the base used, you get soaps with different properties. Lard from hogs differs from tallow from cattle or sheep in that lard contains more oleic acid. Because the salt of a saturated long-chain acid makes a harder, more insoluble soap, the soap formed from tallow is less soluble than that formed from lard. Using different bases to saponify the fats also changes the properties of the soap. Sodium hydroxide produces a harder soap than that made from potassium hydroxide.

Soaps that we commonly use for bathing (toilet soaps) have been carefully washed free of any base remaining from the saponification. Floating soaps, such as Ivory soap, are produced by blowing air into the soap as it solidifies. Scouring soaps have added abrasives, such as fine sand or pumice.

Those of you that live in areas with hard water are probably familiar with a common problem with soaps. Soaps react with divalent cations to form insoluble precipitates we commonly refer to as "soap scum."

Hard water contains divalent metal cations, such as, Fe^{2+}, Ca^{2+}, and Mg^{2+}.

$$2\,RCOO^-Na^+ + Mg^{2+}(aq) \longrightarrow (RCOO^-)_2\,Mg^{2+} + 2\,Na^+(aq)$$

Water soluble $\qquad\qquad\qquad$ Precipitate

An ion exchange column can be used to soften water. The column contains synthetic resin that holds sodium ions (or potassium ions). As hard water passes through the resin, the hard water ions are attracted to the beads and exchange themselves for the sodium or potassium ions. This process produces water that contains sodium or potassium ions instead of the divalent metal ions.

MATERIALS

Equipment

- ✓ 25-mL Erlenmeyer flask
- ✓ 50-mL beaker
- ✓ sand bath
- ✓ hot plate
- ✓ thermal gloves
- ✓ thermometer
- ✓ crystallizing dish
- ✓ Büchner funnel
- ✓ side-arm flask
- ✓ 10-mL graduated cylinder
- ✓ test tubes
- ✓ cork

Chemicals

- ✓ 3 M NaOH in 50:50 water:ethanol
- ✓ lard
- ✓ 50:50 water:ethanol solution
- ✓ saturated sodium chloride solution
- ✓ 4% aqueous $CaCl_2$
- ✓ sodium phosphate (Na_3PO_4)

PROCEDURE

Part A: Synthesis of soap

1. Place about 0.75 g of lard in a 25-mL Erlenmeyer flask.

2. Add 6 mL of sodium hydroxide solution to the flask.

3. Heat the mixture in a sand bath of about 120°C. Place an inverted 50-mL beaker over the neck of the flask to reduce evaporation. Swirl the Erlenmeyer flask every few minutes.

CAUTION! Use thermal gloves when swirling flask.

4. Within about 20 minutes, the soap generally begins to precipitate from the boiling mixture. If it appears that some of the alcohol and water is evaporating from the flask, you should add up to 1 mL of a 1:1 water:alcohol mixture to replace the solvent that is lost. Heat the mixture for a total of 25 minutes in the sand bath.

5. Remove the flask from the sand bath. Add 10 mL of saturated sodium chloride solution. Swirl the mixture while cooling the flask in an ice-water bath.

6. Collect the prepared soap on a Büchner funnel by vacuum filtration on filter paper. Don't forget to insert the filter paper!

7. Wash the soap with two 5-mL portions of ice-cold distilled water to remove any excess sodium hydroxide. Continue to draw air through the filter for a few minutes to partially dry the product.

8. Test your soap while it is still damp.

Part B: Testing your soap

1. After rinsing your product, remove a pea-sized piece of soap from the filter paper and place it in a clean test tube.

2. Add 3 mL of distilled water, cork the test tube, and shake the mixture vigorously for about 15 seconds. After about 30 seconds, observe the level of the foam.

3. Add 2 drops of 4% aqueous calcium chloride to the soap mixture. Shake the mixture for 15 seconds and allow it to stand for 30 seconds. Observe the effect of the calcium chloride on the foam.

4. Add 0.5 g of sodium phosphate and shake the mixture again for 15 seconds. After 30 seconds, again observe the results.

5. Using a clean test tube, test a pea-sized piece of soap with tap water.

NAME _____

RESULTS

1. Describe the appearance of your product, soap.

2. What was the height of the foam formed in distilled water?

3. What was the effect of adding calcium chloride solution to your soap solution?

4. Describe what happened when you added sodium phosphate to the solution.

5. Compare the above results with the test with tap water. Do you believe you have hard or soft water in your area? Justify your answer.

Experiment 31

Titration of Amino Acids

Discovering the effect of pH on amino acids leads to an understanding of protein structure.

Alpha amino acids are the building blocks of proteins. All amino acids contain two functional groups, a primary amine group, $-NH_2$, and a carboxylic acid group, $-COOH$. In addition, there is an R group that is different for each amino acid. The symbol R is used here to represent an organic group.

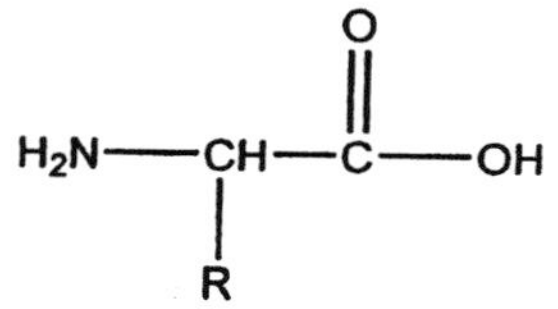

All proteins consist of various combinations of the approximately 20 amino acids.

In physiological systems where the pH is near neutrality (physiological pH is generally close to 7.2), the amino group of an amino acid will be protonated ($-NH_3^+$) and the carboxylic acid group will be unprotonated ($-COO^-$). This form is called the **zwitterion**, a molecule that contains both a positive and a negative charge and thus overall is neutral.

The zwitterion form

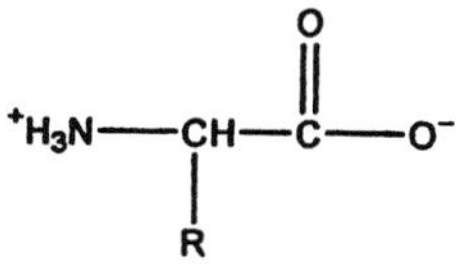

In strongly acidic solutions (pH < 3), the carboxylic acid group will be protonated; whereas in strongly basic solutions (pH > 11), both the carboxylic acid group and the amino group will be unprotonated. Therefore, the overall charge on the amino acid is pH dependent.

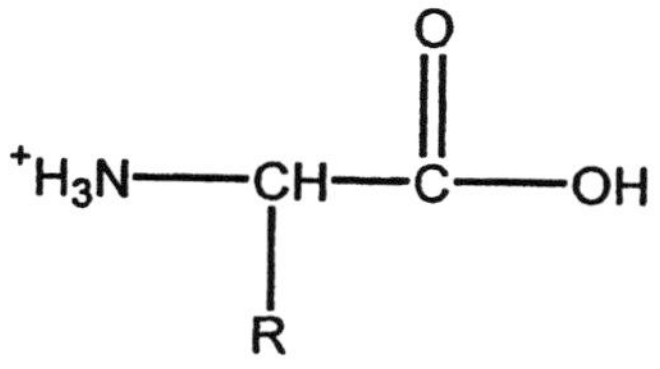

Acidic form 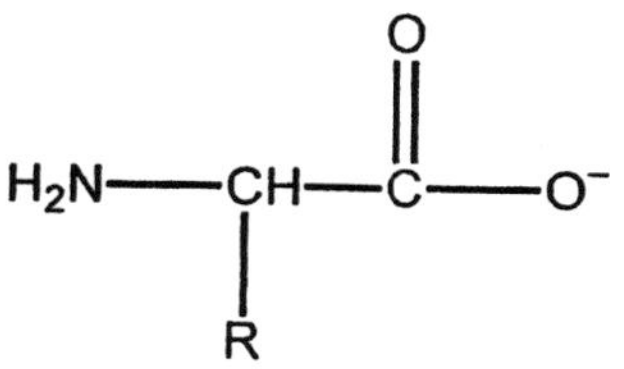   Basic form

215

The acid-base behavior of amino acids is best described by Brønsted theory of acids and bases. A **simple amino acid** is a diprotic acid when in its fully protonated form; it can donate two protons, successively, during its complete titration with a base. The titration with NaOH is a two-stage titration represented below.

$$^+H_3NCHRCOOH + OH^- \rightarrow {^+H_3}NCHRCOO^- + H_2O$$

$$^+H_3NCHRCOO^- + OH^- \rightarrow H_2NCHRCOO^- + H_2O$$

A simple amino acid is one that does not have an acid or base group in the R group.

The titration curve for a simple amino acid will be biphasic (see diagram below). There will be two separate flat portions (called **legs**) on the titration curve. The midpoint of the first leg (B) is where the total concentration of the amino acid is half in the acidic form and half in the zwitterion form. The point of inflection (C) occurs when all of the original amino acid is in the zwitterion form. The actual pH at which this occurs is called the **isoelectric point**, and is represented by the symbol pI. In the pH titration of an amino acid with a nonionizable R group, the equivalence point occurs at the pI of the amino acid. Since the amino acid contains two units (or equivalents) of acid the titration can continue. At the midpoint of the second leg (D), half the amino acid is in the zwitterion form and half is in the basic form.

The **equivalence point** is the stage of the titration at which the volume of titrant (in this case, hydroxide ion) added supplies the number of moles exactly equal to the number of moles of acid being titrated.

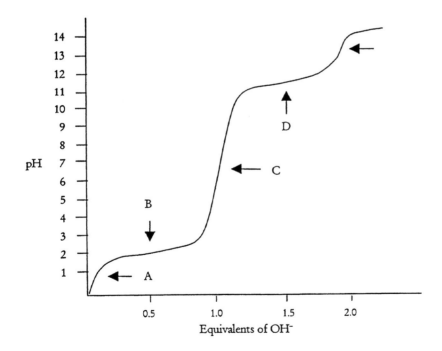

A: 100% $^+H_3NCHRCOOH$
B: 50% $^+H_3NCHRCOOH$ and 50% $^+H_3NCHRCOO^-$
C: 100% $^+H_3NCHRCOO^-$
D: 50% $^+H_3NCHRCOO^-$ and 50% $H_2NCHRCOO^-$
E: 100% $H_2NCHRCOO^-$

The $pK_a = -\log K_a$

Acid dissociation constant (K_a) for the dissociation of acid HA:
$K_a = [H^+][A^-]/[HA]$

Points B and D have a special significance. For example, at point B, the concentration of the acidic form of the amino acid is equal to the concentration of the basic form (50% $^+H_3NCHRCOOH$: 50% $^+H_3NCHRCOO^-$). The pH at this point is equal to the pK_a of the carboxylic acid.

This can be shown by the Henderson-Hasselbalch equation, shown below, which is derived from rearranging the formula for the acid dissociation constant (K_a).

$$pH = pK_a + \log \frac{[base]}{[acid]}$$

Point B is called the pK_{acid} (pK for the carboxylic acid group). Because this point is where half the acid group has been titrated, the concentrations of acid and base are equal and therefore the second term in the equation $\left(\log \frac{[base]}{[acid]}\right)$ is equal to one (1).

$$pH = pK_a + \log 1 = pK_a$$

The log of 1 is zero; therefore, at this particular point (B), $pH = pK_a$.

The same argument can be made for point D. The pH at point (D) is equal to the pK_{amine}.

In this experiment, you will titrate an unknown amino acid and determine its pI, pK_{acid}, and pK_{amine}. These values will be used to determine the identity of the unknown amino acid.

MATERIALS

Equipment

- ✓ 10-mL buret
- ✓ 10-mL pipet
- ✓ magnetic stirring bar
- ✓ pH meter
- ✓ 100- and 250-mL beakers
- ✓ pipet bulb
- ✓ 150-mL beakers (3)
- ✓ 100-mL graduated cylinder
- ✓ 25-mL pipet

Chemicals

- ✓ NaOH solution (~0.3 M)
- ✓ deionized water
- ✓ amino acid unknowns

PROCEDURE

1. Soak the pH electrode in deionized water while preparing the amino acid sample and setting up the buret.

2. Thoroughly rinse the buret with tap water and then with deionized water. The final deionized water rinse should drain evenly from the inside surfaces of the buret and leave no droplets of water behind. Repeat the rinsing procedure until the buret drains cleanly.

3. Obtain approximately 40 mL of NaOH solution in a dry beaker. Record the concentration. Rinse and then fill the buret. Remove any air bubbles and note the starting volume.

4. Obtain approximately 35 mL of an unknown amino acid hydrochloride salt solution. Rinse the 10-mL pipet with the deionized water and then rinse with two small portions of the amino acid solution. Take one clean 150-mL beaker and pipet 10 mL of amino acid solution into the beaker. Add 25 mL of deionized water.

5. Place the pH electrode assembly and a magnetic stirring bar into the beaker. If the electrode is not properly immersed, the pH reading will be erratic.

   Make sure your pH meter has been calibrated before beginning the titration.

6. Titrate the amino acid solution with the NaOH from the buret. This titration should be done at least three times. The first run is done by adding the NaOH at ~1-mL intervals until you are just past the first end point (the "dry run"). Subsequent runs can now be done at ~0.5 mL intervals *until just before each end point and then add titrant in small aliquots until just after each end point*. Record the pH and volume added for each addition.

   The dry run will give you some idea of where the end points are so that you can determine when to add NaOH dropwise.

7. Continue until the pH reaches about 12.5 or until 2.5 equivalents of NaOH have been added.

8. Determine the unknown amino acid from the following possibilities.

proline	$pK_{acid} = 2.0$	$pK_{amine} = 10.6$
glycine	$pK_{acid} = 2.3$	$pK_{amine} = 9.6$
asparagine	$pK_{acid} = 2.0$	$pK_{amine} = 8.8$
serine	$pK_{acid} = 2.2$	$pK_{amine} = 9.15$

DATA

Concentration of NaOH solution: _____

Unknown ID: _____

Dry run titration (addition in 1 mL aliquots)

mL NaOH added	pH	mL NaOH added	pH

Determine the volumes where you need to begin adding more slowly.

Endpoint 1 _____   Endpoint 2 _____

Titration 1

mL NaOH added	pH	mL NaOH added	pH

Titration 2

mL NaOH added	pH	mL NaOH added	pH

RESULTS

1. Prepare a graph of your results, plotting mL of NaOH versus pH.

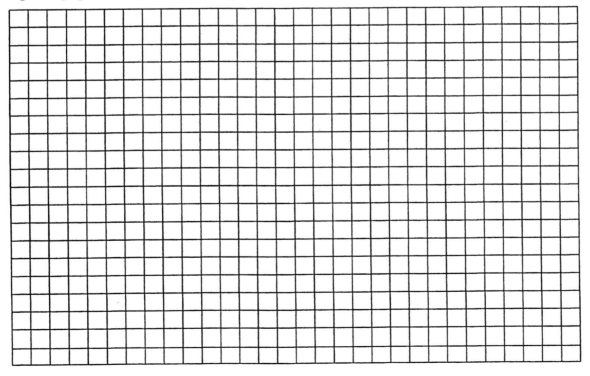

2. The pK_{acid} is the midpoint on the first leg and the pK_{amine} is the midpoint of the second leg. Determine the pK_{acid} and pK_{amine} for your amino acid.

   pK_{acid} _____   pK_{amine} _____

3. The pI can also be calculated by averaging the two pK's. Determine the pI.

   (pK_{acid} + pK_{amine})/2 = pI

4. On the basis of the pK_{acid} and pK_{amine}, identify the amino acid you titrated. Explain your selection.

5. For your amino acid, draw the structures of the species present at pK_{acid}, pI, and pK_{amine}.

Experiment 32

Determination of the Molecular Weight of an Unknown Protein

Gel filtration is used to determine the molecular weight of proteins.

Gel Filtration

One of the major goals in the modern biochemistry laboratory is the isolation of specific proteins from complex mixtures. Such isolation procedures frequently involve chromatography. In chromatography, there is always a stationary phase and a mobile phase, and the separation of compounds is based on the fact that different compounds can distribute themselves to various extents between these different phases. The mobile phase flows over the stationary phase and carries the sample to be separated along with it. The various components in the sample interact with the stationary phase to different extents based on various physical and/or chemical properties. The components that interact more strongly with the stationary phase are carried along more slowly by the mobile phase than are those that interact less strongly. The different mobilities of the components, which in turn are based on the various extents of interaction with the stationary phase, form the basis of the separation.

In gel filtration, the stationary phase consists of gel beads that are permeated with pores. Separation by gel filtration depends on the different abilities of sample molecules to enter these pores. Very large molecules never enter the stationary phase, so they move through the chromatographic bed the fastest. Above a certain size (the **exclusion limit**), all molecules move equally fast and no separation occurs. Very small molecules easily enter the stationary phase (they are included), so they move through the bed slowest. Below a certain size, all molecules move equally slowly and, again, no separation occurs. Between these extremes, molecules enter the stationary phase to various extents, depending on their size and shape. Within this range, called the **fractionation range**, different molecules move at different rates and separation occurs.

The two most common polymers used to form the gel are polyacrylamide and dextran. The polymers are cross-linked to themselves to form small beads, which swell upon addition of water to form structures that resemble microscopic porous sponges. The cross-linked structure of these polymers produces pores in the materials. The pore size can be selected for a desired value, which depends on the extent of the cross-linking. Cross-linked dextrans have the smallest pores, with exclusion limits ranging from less than 1000 to approximately 250,000. Exclusion limits for other types of media range from 100,000 to 100,000,000. Beads with large pores are used for separating large protein molecules, whereas small proteins and peptides are separated on gels with small pores.

To separate two proteins of different size by gel filtration, a sample containing a mixture of proteins is carefully introduced to the top of the gel bed. Buffer (the mobile phase) is added to the top of the column and the proteins enter the gel. The buffer is allowed to drip from the column into a series of test tubes and the amounts and types of the proteins in each tube (**fraction**) are determined. This process is shown in the marginal figure.

The Stages of Separation of Two Proteins on a Gel Filtration Column

1. To prepare the gel bed, the porous hydrated beads (large open circles) are packed into a column.

2. A sample containing a mixture of a large protein (large black dots) and a small protein (small black dots) is added to the top of the gel bed.

3. To separate the proteins, the buffer is applied to the top of the column and the proteins enter the gel bed. The smaller proteins penetrate the gel beads while the larger proteins are excluded from the gel beads and thus move rapidly down the column.

4. The two proteins have separated and can be collected in different fractions of the column eluate.

In gel filtration, the molecular weight of a molecule is determined by measuring elution volume (V_e). The **elution volume** of a molecule is defined as the volume of buffer required to elute the molecule from the column. To compensate for any variations in the flow rate, the elution volume is compared with the void volume (V_o). The **void volume** is the elution volume for a substance that is completely excluded from the gel. A common material for determining the void volume is blue dextran, a very large colored polysaccharide with a molecular weight of about 2,000,000. Blue dextran does not enter the pores in the gel beads because of its large size, and therefore it flows only between the beads during its journey down the column. The ratio V_e/V_o is calculated and plotted versus the log of the molecular weight. As an example for calculating this ratio, consider the following data for three molecules: blue dextran, the protein myoglobin, and phenol red are placed on a column. Their elution volumes are 5 mL, 15 mL, and 20 mL, respectively. The void volume is the volume required to elute the blue dextran, which is 5 mL. The V_e/V_o for blue dextran is 1; the V_e/V_o for myoglobin is 15 mL/5 mL = 3; and for phenol red, the V_e/V_o is 20 mL/5 mL = 4. This information can then be used to determine the molecular weight (MW) of a protein by comparing its elution profile (V_e/V_o) with the elution profiles (V_e/V_o's) of standard proteins of known molecular weights. In practice, this determination is done by preparing a standard graph of the *logarithms* of the known molecular weights plotted against their respective V_e/V_o's. A linear relationship exists between the logarithms of the molecular weights of the proteins and their respective V_e/V_o's. Consider the following example: Five molecules of known molecular weight (let's call them A, B, C, D, and E) and one molecule of unknown molecular weight (let's call it F) are place on a gel filtration column. The following data was collected:

Molecule	MW	log MW	Color	V_e	V_e/V_o
A (blue dextran)	2,000,000	6.3	blue	3.0 mL	1
B	400,000	5.6	brown	4.5 mL	1.5
C	80,000	4.9	orange	6.0 mL	2
D	3200	3.5	pink	9.0 mL	3
E	630	2.8	red	10.5 mL	3.5
F	?	?	yellow	7.5 mL	2.5

Graphing the log of the molecular weight versus the V_e/V_o's for the known proteins gives the following graph. From the graph, the log of the molecular weight of the unknown molecule can be extrapolated to be 4.2, which gives a molecular weight of $10^{4.2} = 16,000$.

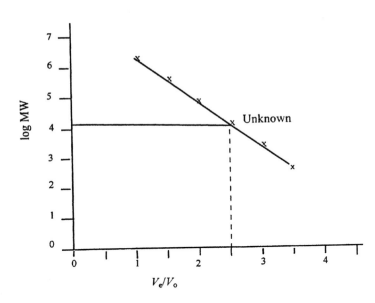

In this experiment, you will utilize gel filtration to determine the molecular weight of an unknown orange-colored protein. To accomplish this, you will use the following molecules as standards:

Molecule	Chemical nature	Color	Molecular weight
blue dextran	polysaccharide	blue	2,000,000
cow hemoglobin	protein	brown	64,000
cyanocobalamin	vitamin (B_{12})	pink	1,355
phenol red	organic dye	red-orange	360

MATERIALS

Equipment

- plastic column with stopcock and reservoir
- 24-well plate
- calibrated micropipet
- transfer pipets
- ring stand and clamps
- stopwatch
- small beakers (2)
- 10-mL graduated cylinder

Chemicals

- Sephadex 150 superfine
- column buffer: 20 mM Tris HCl, 0.1 M NaCl, pH 7.4
- protein and dye solutions

PROCEDURE

Part A: Packing the column bed

1. Secure the column to a ring stand in a vertical position with the reservoir on top and the stopcock at the bottom.

2. The gel has been preswelled by allowing it to stand overnight in an excess of water. Using a disposable pipet to transfer 10 mL of the preswelled gel slurry from the bottom of its container to a small beaker.

3. Obtain approximately 50 mL of column buffer in a small beaker. Fill the column to about 20% of its length with the column buffer. Thoroughly mix the gel slurry to make sure that it is uniformly suspended, and add *part* of the slurry to fill the column. (*Note:* The slurry can be added to the column with a disposable pipet or by pouring carefully from a small beaker or cylinder.)

4. Allow a portion of the gel to settle. As the gel is settling, you will observe the formation two layers: a bottom opaque layer of gel and a clear top layer of buffer. A 2-cm gel bed will form in a few minutes. Place a small beaker under the column and open the stopcock. Buffer will flow out of the column into the beaker, and the gel will slowly settle.

5. Add more gel slurry in small portions until the gel bed is 3 to 7 cm from the top of the column tube. Add column buffer until the column is filled.

6. Allow buffer to flow through the column and drain out the bottom for several minutes. This will "pack" your column. As your column packs down, you may decide you need to add more gel. To do this, you'll want to pipet the column buffer off the top of the column, making sure to leave enough buffer so that your column doesn't dry out. Add more gel slurry and fill the column again with column buffer.

7. Once you have your column packed, close the stopcock and proceed to the next step: checking the column flow rate.

Part B: Determination of the column flow rate

The optimum rate of buffer flow is roughly 1 to 1.7 mL/10 minutes. This value is subject to column variation and must be determined experimentally for your column.

1. Place a dry 10-mL graduated cylinder under the column to catch the buffer that will flow through.

2. Open the stopcock and allow buffer to collect in the graduated cylinder for *exactly* 10 minutes.

3. Record the flow rate of your column as follows.

    Flow rate

    _____ mL / 10 minute   Desired rate is 1 to 1.7 mL/10 minutes.

    _____ mL / 3 minutes*

* This value will be used as the volume of each column fraction.

Part C: Sample application and separation

1. Stop solution from flowing out of the column by closing the stopcock on the column. Using a disposable pipet, remove nearly all the column buffer from the top. Be very careful not to disturb the gel in the column *and do not let the column run dry!*

    > **CAUTION!**
    > Do not let the column run dry. If the column runs dry, you must repack the column.

2. Using calibrated *micro*pipet place 0.1 mL of the sample mixture into a microfuge tube.

3. Transfer the 0.1-mL sample to the column. To do this, place the tip of the pipet about 1 to 2 mm above the gel bed and carefully layer the sample onto the upper bed surface. When adding the sample to the column, be careful to disturb the gel as little as possible.

4. Open the stopcock and let the sample drain into the bed. Add column buffer dropwise to ensure the column does not dry out while the sample is draining into the column.

5. Once the sample is on the column, carefully fill the column with column buffer. Again, disturb the gel as little as possible. (*Note:* If the sample is not on the column, it will dissolve in the buffer above the column and thereby color the buffer. This error will give you a poor separation.)

6. Place the 24-well collection plate under the column and begin to collect the first fraction.

7. After 3 minutes, move the collection plate so that the second well is under the column and collect the second fraction for 3 minutes. (*Note:* If your flow rate is significantly faster than 1 to 1.7 mL/10 minutes, collect 2 minute fractions.) *Do not let the column run dry!*

8. Repeat the above process until the five colored substances have been eluted from the column. During collection, watch the separation of the substances as they progress down the column.

9. Examine the column fractions in the collection plate and identify those fractions that contain the highest concentrations of blue dextran (blue), bovine hemoglobin (brown), the unknown (orange), cyanocobalamin (pink), and phenol red (orange-red). These can be seen most clearly when the collection plate is placed on a sheet of white paper. Record this information in the data section.

10. Determine the identity of the unknown from the following possibilities.

Possible unknowns	Molecular weights
cytochrome C	12,400
carbonic anhydrase	29,000
ovalbumin	45,000
phosphorylase B	97,400
alcohol dehydrogenase	150,000

DATA AND RESULTS

Flow rate: _____

Elution time is determined by multiplying the fraction number by the time it took to collect each fraction. If 3 minutes was used to collect each fraction, multiply the fraction number by 3 to determine the elution time. The elution volume is determined by multiplying the elution time by the flow rate.

Sample	Fraction number	Elution time	Elution volume
blue dextran			
bovine hemoglobin			
unknown			
cyanocobalamin			
phenol red			

1. Calculate the V_e/V_o for each molecule.

    Void volume (blue dextran): _____

Sample	V_e/V_o
bovine hemoglobin	
unknown	
cyanocobalamin	
phenol red	

2. Determine the log of the molecular weight for each known molecule.

Sample	Log MW
bovine hemoglobin	
unknown	
cyanocobalamin	
phenol red	

3. Plot the log of the molecular weight versus the V_e/V_o's for the known molecules.

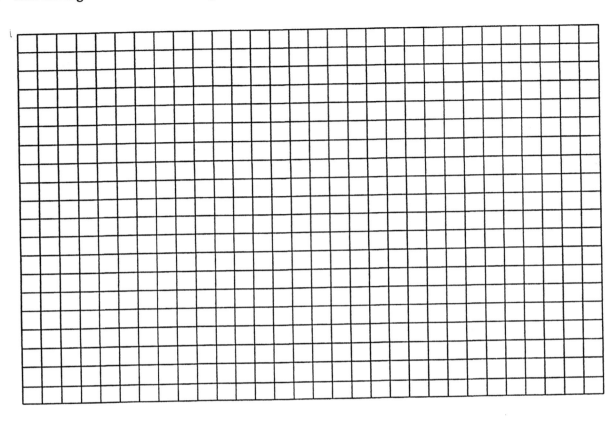

4. From the graph, extrapolate log MW of the unknown compound, using its V_e/V_o.

   Log MW of unknown compound: _____

5. Use your calculator to determine the MW of the unknown.

   MW unknown: _____

6. Determine the identity of the unknown from the possibilities given above. Explain your choice.

Electrophoresis of Normal and Sickle Cell Hemoglobin

A small change in amino acid sequence can result in a dramatic change in protein shape.

<small>The overall structure of hemoglobin is written as $\alpha_2\beta_2$ in Greek notation.</small>

Red blood cells, or erythrocytes, carry the protein hemoglobin (Hb). The function of hemoglobin is to transport oxygen from the lungs to the tissues. Hemoglobin is a globular protein made up of four subunits, two alpha (α) chains and two beta (β) chains. Each subunit also contains a heme group containing iron, iron being the site of O_2 binding.

Many changes in the structure of hemoglobin have arisen by mutations in the human population. About 1 person in 100 has a mutant hemoglobin gene, and these individuals have abnormal hemoglobin molecules in their blood. Not all mutant hemoglobin's have clinical manifestations. For example, Hb E occurs in up to 10% of the Southeast Asia population. In Hb E, the glutamic acid (glu) at position 26 of the β-chain has been replaced by lysine (lys), denoted Glu (26) $\beta \rightarrow$ Lys. This amino acid residue is located on the surface of the protein and has little effect on the stability. However, in a few cases, mutations in hemoglobin can cause serious diseases. One of the most common and serious abnormal hemoglobins is hemoglobin S, which is present in individuals suffering from sickle cell anemia. In Hb S, a single glutamic acid residue on the β chains is replaced by valine (val), Glu (6) $\beta \rightarrow$ Val.

Residues number	4	5	6	7	8	9
Normal Hb	-Thr	-Pro	-**Glu**	-Glu	-Lys	-Ala
Sickle cell Hb	-Thr	-Pro	-**Val**	-Glu	-Lys	-Ala

This single change in the primary sequence causes a marked change in the net charge and conformation of the protein. Glutamic acid is acidic, and under biological conditions it is negatively charged. Valine is a nonpolar amino acid, and its presence in the hemoglobin creates a hydrophobic interaction between the β chains in adjacent hemoglobin molecules. Because of this hydrophobic interaction, the cell becomes deformed and assumes a sickle shape. The sickling of the cell is triggered by low concentrations of oxygen. When Hb S is deoxygenated, the interaction between the chains occurs, the hemoglobin crystallizes in the red blood cells, and the sickle cells are unable to carry oxygen. The sickle cells are also rigid, so they clog capillaries. The body's defense system quickly destroys the defective cells, which leads to a reduced number of erythrocytes; hence, the term sickle cell anemia.

Sickle cell disease is inherited as a recessive gene. When the gene for Hb S is inherited from only one parent, the individual is heterozygous for the condition and is a sickle cell carrier. Although these individuals rarely have severe anemia, half of their circulating hemoglobin is Hb S and half is normal Hb A. When the gene for Hb S is inherited from both parents, the individual is homozygous. In this case, all the hemoglobin is Hb S and the individual suffers from severe anemia.

The gene for Hb S is believed to have originated in the black population in Africa. Hemoglobin S has an important benefit in the African population because it confers resistance to one form of malaria. This malarial parasite cannot thrive in the abnormal red blood cells. Thus, heterozygous carriers will survive malaria in higher numbers than those with homozygous normal or homozygous recessive hemoglobin will. In some parts of Africa, up to 40% of the population have the sickle cell trait. The incidence of sickle cell trait in the African-American population is about 10%, and 1 to 2% of this population has sickle cell anemia.

Abnormal hemoglobin is often detected in the clinical laboratory by electrophoresis. The isoelectric point of normal Hb A is 6.9. At pHs greater than 7, the hemoglobin molecules have a net negative charge. Because glutamic acid residues (negatively charged) are replaced by valine residues (neutral) in the β chains of Hb S, the electrophoretic properties are correspondingly affected. In this experiment, you will compare the electrophoretic behavior of normal and sickle cell hemoglobin.

The isoelectric point is the pH at which the charge on hemoglobin is zero.

MATERIALS

Equipment
- ✓ 100-mL beaker
- ✓ watchglass
- ✓ gel box with leads and power supply
- ✓ 100-mL graduated cylinder
- ✓ stirring rod
- ✓ pipettors, volume range 2 to 20 μL
- ✓ thermal gloves
- ✓ microfuge tubes (4)
- ✓ grease pencil

Chemicals
- ✓ agarose powder
- ✓ 0.01 M sodium tetraborate, pH 9.2 buffer
- ✓ sample loading buffer (30% glycerol, 50 mM EDTA, 0.25% bromophenol blue)
- ✓ normal hemoglobin (1 μg/μL)
- ✓ sickle cell hemoglobin (1 μg/μL)
- ✓ Coomassie Blue gel stain
- ✓ 10% acetic acid solution

PROCEDURE

Part A: Preparation of the agarose gel

CAUTION! Beaker and agarose suspension will be hot. Use thermal gloves.

1. Weigh out 0.6 g of agarose and place in a 100-mL beaker. Add 50 mL of pH 9.2 buffer. Cover your beaker with a watchglass and heat this mixture in a microwave for 2 minutes until all the agarose particles are dissolved. Use the thermal gloves to remove the beaker from the microwave. Caution: HOT!!

2. Cool the solution to about 60°C. Pour approximately 25 mL of slightly cooled agarose solution into the gel tray (just fill the tray up to the edge). Check to make sure that there are no bubbles in the molten agarose. If there are any bubbles in the gel, gently shake the gel bed. Bubbles must be removed before the gel solidifies.

3. Insert the comb at the negative electrode (black) end. Let the gel harden for 10 minutes.

Part B: Preparation of your samples

1. Label the four microfuge tubes so that you can identify your samples. You will be running the following samples:

   normal hemoglobin, two lanes 4 µL and 6 µL

   sickle cell hemoglobin, two lanes 4 µL and 6 µL

2. Prepare your samples so that you have a total volume of 17 µL in each tube. Add the reagents to each microfuge tube in the following order: pH 9.2 buffer first, hemoglobin next, sample loading buffer last.

Tube	Sample	pH 9.2 buffer	hemoglobin	Loading buffer
1	normal	10 µL	4 µL	3 µL
2	normal	10 µL	4 µL	3 µL
3	sickle cell	8 µL	6 µL	3 µL
4	sickle cell	8 µL	6 µL	3 µL

Part C: Electrophoresis

1. Add pH 9.2 buffer to cover the gel by about 1.5 mm (about 125 mL). Gently remove the comb and dams. Put the gel box near the power supply before loading the samples.

2. Load 15 µL of your samples into the wells. You may need to tap the microfuge tubes on the desktop to bring down any drops on the sides of the microfuge tube.

3. When all samples are loaded, close the lid on the gel box. Check that the power supply is turned off and the voltage is turned all the way down. Attach the electrodes to the gel box and power supply, making sure that the red leads connect the + terminals and the black leads connect the − terminals. If the power supply allows variable voltage, adjust the voltage to 100 V.

   CAUTION! The power supply should be turned off and voltage turned down before connecting the electrodes to the power supply.

4. Run the electrophoresis for 60 to 70 minutes or until the purple tracking dye has moved at least 2/3 of the way down the gel.

5. Before removing the gel, turn off the power supply and detach the electric leads. Gently transfer the gel to a container of Coomassie Blue stain. Keep the gel in the stain for at least 30 minutes. At the end of 30 minutes, pour the stain off into another vessel.

   The stain can be reused! Place it into a container to be recycled. Do not throw it away!

6. Add 10% acetic acid solution (destain) to your container and destain for as long as time permits. It is useful to change destain periodically. Simply discard the old acetic acid and pour in fresh solution. The longer you can destain the more prominent the protein bands will appear. You can destain too long and remove the blue stain from the protein completely, so do not destain for more than 60 minutes.

7. When finished, draw a picture of the gel.

NAME_____

RESULTS

Draw the visible bands on the diagram below.

Wells

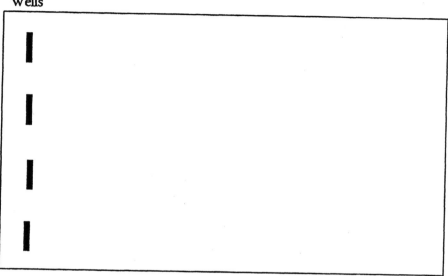

1. Explain the difference in migration of the two hemoglobins with respect to the known mutation in sickle cell hemoglobin.

2. You arrive in lab and a pH 8.0 buffer has mistakenly been supplied instead of the pH 9.2 buffer. Would you expect to see any difference in your results? Why or why not?

Experiment 34

Separation of Albumin from Serum by Affinity Chromatography

Affinity chromatography is an excellent way to purify a protein.

A substantial portion of a biochemist's time is devoted to the extraction and purification of biologically active compounds. This isolation process can be a formidable problem because the compound of interest is just one among thousands of different substances. One additional complication is that generally the compound of interest is present in extremely small amounts.

One technique used for the isolation, purification and initial characterization of a biologically active compound is affinity chromatography. Affinity chromatography is a separation technique that makes use of the compound's unique biological properties. That is, it makes use of a special noncovalent binding affinity between the biological compound and a special molecule, called the **ligand**. Several examples of ligand and protein interactions that can be exploited for protein isolation are given below.

> A **ligand** is an ion or molecule coordinated to a central atom or molecule in a complex.

For purification of	Suggested ligand
insulin receptors	insulin
glycoproteins	carbohydrates
chymotrypsin	4-phenylbutylamine
fucosidase	fucosylamine
lipoproteins (HDL and LDL)	cholesterol

The ligand–protein interactions given in the above list are reasonable. If you want to isolate insulin receptors, you use insulin as the ligand. Or if you want to isolate chymotrypsin, which cleaves peptide bonds on the carboxy side of an aromatic amino acid, it is reasonable to use an aromatic ligand such as 4-phenylbutylamine. Research has also shown that some ligands do not need to be specific and that some enzymes can be isolated with several different ligands. Some proteins can be isolated by using an organic dye as a ligand.

Organic dyes offer a "nonspecific" affinity medium. It is believed that these organic dyes mimic biological substances (substrates, cofactors, and effectors) and thereby produce the ligand–protein binding affinity effect. Because proteins react differently with different dyes, it is often possible to develop an optimized purification protocol by screening different dye resins and eluants.

In affinity chromatography, the ligand is covalently bound to an insoluble matrix, such as agarose, which is placed in a column. The mixture containing the protein of interest is loaded on the affinity column. After the nonbinding protein molecules have passed through the column, the protein of interest is removed from the column by altering the conditions that affect the binding (i.e., pH or salt concentration).

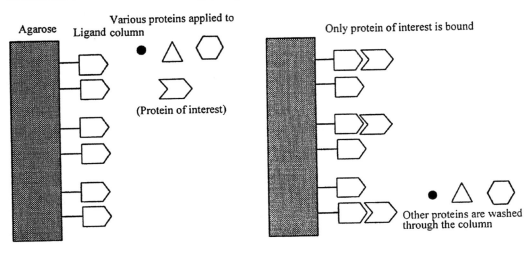

In this experiment, you will use Affi-Gel Blue affinity gel to isolate serum albumin. Affi-Gel Blue affinity gel is a beaded, cross-linked agarose gel with covalently attached Cibacron Blue F3GA dye (see structure on next page). Affi-Gel Blue gel purifies a large range of proteins from widely divergent origins. The blue dye functions as an ionic, hydrophobic, aromatic, or sterically active binding site in various applications. Proteins that interact with Affi-Gel Blue gel can be bound or released with a high degree of specificity by manipulating the composition of the eluant buffers.

The binding of serum albumin to Affi-Gel Blue is very strong, but albumin can be eluted from the column by using high salt concentrations. Because salt is used to remove the albumin from the column, proper adjustment of the sample ionic strength is critical for optimal binding of albumin. To achieve this, the serum sample is applied to a desalting column prior to application to Affi-Gel Blue.

Cibacron Blue coupled to agarose

In desalting, a matrix is chosen for its ability to exclude the larger solutes of interest from the pores of the gel and to retain the smaller contaminants. The Econo-Pac 10DG columns used in the experiment are packed with a matrix that excludes solutes greater than 6000 daltons, allowing them to elute in the void volume. Therefore, desalting occurs between the excluded components larger than 6000 daltons and the included components smaller than 6000 daltons (e.g., salt). You will load serum onto the desalting column, and your serum proteins will be eluted in the void volume. You will then take your desalted serum sample and apply it to an Affi-Gel Blue column, which will separate the albumin from the rest of the serum proteins.

MATERIALS

Equipment

- 50-mL beaker
- 400-mL beaker
- Econo-Pac 10 DG desalting column
- ring stand and clamps
- plastic column
- 10-mL graduated cylinder
- 50-mL Erlenmeyer flask
- test tubes (3)
- transfer pipets

Chemicals

- serum
- Affi-Gel Blue
- buffer A (20 mM phosphate buffer, pH 7.1)
- buffer B (1.4 M NaCl, in 20 mM phosphate buffer, pH 7.1)
- buffer C (2 M guanidine HCl in 20 mM phosphate buffer, pH 7.1)
- bromcresol purple solution

PROCEDURE

Step 1: Affinity Chromatography of Serum

Part A: Desalting of serum sample

1. Obtain an Econo-Pac 10 DG desalting column and clamp the column to a ring stand. Remove the upper cap and pour off the excess liquid off the top of the column. Do not worry about pouring out the column contents. The contents are held in place by a plug of fritted glass that sits on the top of the column bed.

2. Place a small beaker beneath the column to catch the effluent. Add 20 mL of deionized water to the column and then remove the yellow cap from the tip to start the column flowing.

3. Allow the deionized water to drain to the fritted glass plug. The column will not run dry; flow will stop when the deionized water level reaches the plug. Discard the 20 mL of eluant.

4. Measure 1.0 mL of serum and then add the serum sample to the column.

5. Allow the entire sample to enter the column. Add 2.0 mL of deionized water. Allow the water to enter the column and discard the eluant.

6. Add 1.5 mL of deionized water to elute your serum sample. Collect this fraction in a 10-mL graduated cylinder. Save 1 mL of this sample in a test tube labeled "serum."

Part B: Separation of albumin from serum by affinity chromatography

1. Obtain a chromatography column and clamp it to a ring stand.

2. Obtain ~8 mL of Affi-Gel Blue gel solution and use a transfer pipet to add the gel to the chromatography column with its stopcock closed. Place a small beaker under the end of the column and open the stopcock to start the column flowing. Fill the column with Buffer A and continue to rinse the column with approximately 10 mL of Buffer A solution. Let the buffer drain down to the level of the gel bed. Unlike the desalting column, you must take care to not let the column run dry!

   CAUTION!
   Don't let the column run dry!

3. Apply 1.5 mL of your desalted serum sample to the column.

4. Let the desalted serum sample drain down to the level of the gel bed and then add 10 mL of Buffer A. The eluant from this step contains the serum proteins minus most of the albumin. Collect this 10 mL in a test tube labeled "serum – albumin."

5. Once Buffer A has drained down to the level of the gel bed, elute the albumin by adding 10 mL of Buffer B. Collect this 10 mL in a test tube labeled "albumin."

6. Regenerate the column with 10 mL of Buffer C and place the Affi-Gel Blue gel in the container labeled "regenerated Affi-Gel Blue."

Part C: Testing the samples for the presence of albumin

1. Take 3 test tubes and label A, B, and C. In test tube A put 1 mL of sample eluted with Buffer A. In test tube B put 1 mL of sample eluted with Buffer B. In test tube C, place 1 mL of the original buffer B.

2. Add 1 mL of bromcresol purple solution to each test tube. Bromcresol purple forms a purple complex with albumin. The presence of albumin is indicated by any difference in color from the blank, test tube C.

Step 2: Analysis of Separation by Electrophoresis

The results from the test with bromcresol purple can sometimes be unclear. If there is more albumin present than possible binding sites on the column, you can see albumin present in the test tube A, which should be serum minus albumin. A way to better visualize the sample is by gel electrophoresis. In gel electrophoresis, the proteins of the sample are separated based on their size and charge. The proteins can then be visualized by using a stain the binds to the proteins in the gel.

How can we use this to analyze the separation? You will run three samples: one sample that contains the original desalted serum; another sample that contains the serum proteins minus albumin; and, finally, the sample that contains only albumin. By comparing these three samples, we can look at the efficiency of removal of the albumin from serum. The first sample, desalted serum, gives you the separation pattern for all the proteins in the gel. Because albumin is a major protein component in serum, you will see a very dense band, much larger and denser then any other, for albumin. In the second sample, you will see all the bands that were in the first sample, but the albumin band should be missing or very faint. If the separation was successful, the third sample should only contain albumin, thus you should see only one band that migrates in the gel the same distance as the albumin in the first sample.

You will run two sets of the three samples. The sample will differ in their concentrations. This increase of concentration will increase the probability of seeing the protein bands.

MATERIALS

Equipment
- ✓ 100-mL beaker
- ✓ watchglass
- ✓ gel box with leads and power supply
- ✓ 100-mL graduated cylinder
- ✓ stirring rod
- ✓ pipettors, volume range 2 to 20 µL
- ✓ thermal gloves
- ✓ microfuge tubes (6)
- ✓ grease pencil

Chemicals
- ✓ agarose powder, protein grade
- ✓ 0.01 M sodium tetraborate, pH 9.2 buffer
- ✓ sample loading buffer (30% glycerol, 50 mM EDTA, 0.25% bromophenol blue)
- ✓ sample from Step 1
- ✓ Coomassie Blue gel stain
- ✓ 10% acetic acid solution

PROCEDURE

Part A: Preparation of the agarose gel

1. Weigh out 0.6 g of agarose and place in a 100-mL beaker. Add 50 mL of pH 9.2 buffer. Cover your beaker with a watchglass and heat this mixture in a microwave for 2 minutes until all the agarose particles are dissolved. Use the thermal gloves to remove the beaker from the microwave. Caution: HOT!!

2. Cool the solution to about 60°C. Pour approximately 25 mL of slightly cooled agarose solution into the gel tray (just fill the tray up to the edge). Check to make sure that there are no bubbles in the molten agarose. If there are any bubbles in the gel, gently shake the gel bed. Bubbles must be removed before the gel solidifies.

3. Insert the comb at the negative electrode (black) end. Let the gel harden for 10 minutes.

> **CAUTION!**
> Beaker and agarose suspension will be hot. Use thermal gloves.

Part B: Preparation of your samples

1. Label the six microfuge tubes so that you can identify your samples. You will be running the following samples:

   desalted serum sample, two lanes 4 µL and 6 µL

   serum minus albumin, two lanes 4 µL and 6 µL

   albumin, two lanes 4 µL and 6 µL

2. Prepare your samples so that you have a total volume of 17 µL in each tube. Add the reagents to each microfuge tube in the following order: pH 9.2 buffer first, hemoglobin next, sample loading buffer last.

Tube	Sample	pH 9.2 buffer	hemoglobin	Loading buffer
1	Desalted serum	10 µL	4 µL	3 µL
2	Serum – albumin	10 µL	4 µL	3 µL
3	albumin	10 µL	4 µL	3 µL
4	Desalted serum	8 µL	6 µL	3 µL
5	Serum – albumin	8 µL	6 µL	3 µL
6	albumin	8 µL	6 µL	3 µL

Part C: Electrophoresis

1. Add pH 9.2 buffer to cover the gel by about 1.5 mm (about 125 mL). Gently remove the comb and dams. Put the gel box near the power supply before loading the samples.

2. Load 15 µL of your samples into the wells. You may need to tap the microfuge tubes on the desktop to bring down any drops on the sides of the microfuge tube.

3. When all samples are loaded, close the lid on the gel box. Check that the power supply is turned off and the voltage is turned all the way down. Attach the electrodes to the gel box and power supply, making sure that the red leads connect the + terminals and the black leads connect the − terminals. If the power supply allows variable voltage, adjust the voltage to 100 V.

CAUTION!
The power supply should be turned off and voltage turned down before connecting the electrodes to the power supply.

4. Run the electrophoresis for 60 to 70 minutes or until the purple tracking dye has moved at least 2/3 of the way down the gel. The longer you let the gel run the better will be your protein separation.

The stain can be reused! Place it into a container to be recycled. Do not throw it away!

5. Before removing the gel, turn off the power supply and detach the electric leads. Gently transfer the gel to a container of Coomassie Blue stain. Keep the gel in the stain for at least 30 minutes. At the end of 30 minutes, pour the stain off into another vessel.

6. Add 10% acetic acid solution (destain) to your container and destain for as long as time permits. It is useful to change destain periodically. Simply discard the old acetic acid and pour in fresh solution. The longer you can destain the more prominent the protein bands will appear. You can destain too long and remove the blue stain from the protein completely, so do not destain for more than 60 minutes.

7. When finished, draw a picture of the gel.

NAME_____

DATA

Testing the samples for the presence of albumin

Test tube: sample	Color
A: Desalted serum	
B: Serum – albumin	
C: Albumin	

Analysis of separation by electrophoresis

Draw the visible bands on the diagram below.

Wells

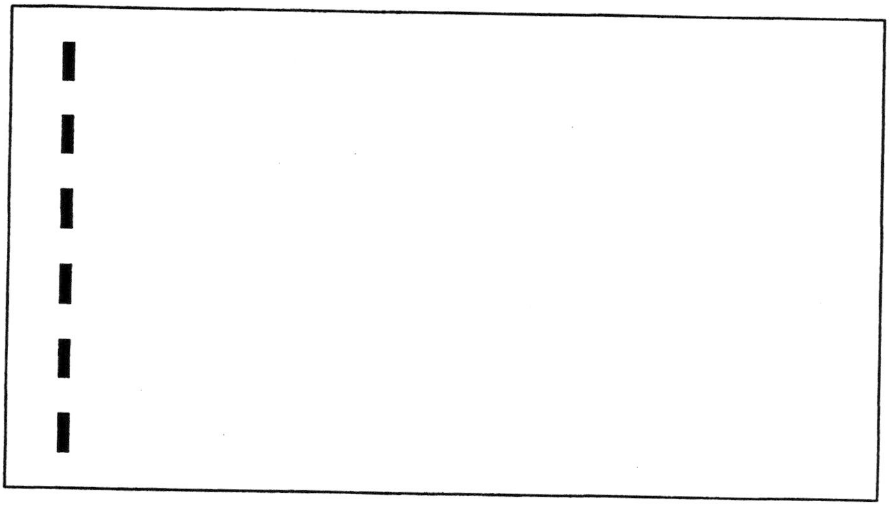

RESULTS

1. The test for the removal of albumin is qualitative. What conclusions can you make from the results of the bromcresol blue test?

2. On the basis of the electrophoresis, comment on the quality of the separation of albumin from serum.

QUESTIONS

1. Would the affinity column have been effective if the serum had been loaded onto the column without desalting? Explain.

2. Because a high salt concentration is used to elute the albumin from the column, speculate on the type of interaction that exists between albumin and Cibacron Blue dye.

Analysis of Wheat Germ Acid Phosphatase

Enzymes allow normally slow reactions to happen quickly.

An **enzyme** is a protein catalyst. Many reactions that take place in the human body are very slow. For example, you can keep a mixture of table sugar (sucrose) and water for a reasonably long period. Yet, when you eat table sugar, it is easily metabolized into its components, glucose and fructose. It is the presence of enzymes in your body that allows you to rapidly carry out reactions that normally take minutes, hours, or perhaps days. The simplest way to represent an enzyme-catalyzed reaction would be as follows:

A catalyst is a substance that increases the rate of a chemical reaction but is not itself changed in the process.

$$\text{Substrate (S)} \xrightarrow{\text{enzyme}} \text{product (P)}$$

The velocity or rate of the reaction can be either determined by measuring the decrease in substrate concentration, $-\Delta[S]$, with time, Δt, or, more commonly, by measuring the rate of appearance of product, $\Delta[P]$, with time:

$$\text{Rate} = -\Delta[S]/\Delta t = \Delta[P]/\Delta t$$

The rate of a reaction is an important property of an enzyme. Enzymes can increase the rate of a chemical reaction 10^3 to 10^{10} fold. Factors that can effect the rate of an enzyme-catalyzed reaction are enzyme concentration, temperature, and pH. Temperature and pH will affect the enzyme itself; each enzyme has an optimal pH and temperature range. At temperatures or pH values outside the optimum, the enzyme can be denatured and inactivated. Therefore, it is important that these two factors be held constant during the measurement of the rate of the enzyme-catalyzed reaction. The rate of an enzyme-catalyzed reaction will be proportional to the concentration of substrate, up to a point. At high substrate concentrations, the enzyme will be saturated—that is, it is going as fast as it possibly can. At this point, added substrate will not increase the rate of reaction.

To determine the rate of an enzyme-catalyzed reaction, one uses a fixed concentration of enzyme and substrate. During the early part of the reaction, the amount of product formed increases linearly with time. However, in the latter part of the reaction, the rate of product appearance diminishes to a point where product is no longer formed. This is illustrated graphically below. There are a number of reasons for the decline in reaction rate with time. These include the depletion of substrate or the breakdown (denaturation) of the enzyme. Thus, the rate or velocity of the reaction is determined during the early time interval when the amount of product is increasing linearly with time. The rate is determined from the slope of a straight line tangent to the beginning of the curve. This rate is called the initial velocity of the reaction (v_0). The v_0 in the illustration below is approximately 15 nanomoles (nmol) of product formed per minute.

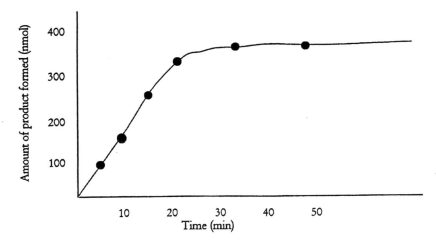

Wheat germ acid phosphatase catalyzes the hydrolysis of phosphate groups from macromolecules and smaller molecules that are stored in the wheat seed. The growing wheat embryo uses the freed phosphate in germination and growth. In this experiment, you will measure the velocity of the reaction catalyzed by purified acid phosphatase. Nitrophenyl phosphate, a colorless compound, will be used as a substrate in the experiment. The hydrolysis products are nitrophenol and phosphate (see below).

Under alkaline conditions, nitrophenol is converted to a nitrophenoxide ion, which is yellow. Using a spectrophotometer, the concentration of product (nitrophenol) can be measured by the increase in absorbance due to the yellow color.

The addition of base at the end of the reaction serves two purposes. First, it converts the product into a colored substance that we can use to measure concentration. Second, it stops the reaction by denaturing the enzyme. In this way, we can determine the concentration of product formed at specific time intervals by simply adding base.

There are two parts to this experiment. First, you will use acid phosphatase to catalyze the hydrolysis of nitrophenyl phosphate. The reaction will be stopped by the addition of base at various time intervals. The concentration of nitrophenol can be determined by measuring the absorbance and extrapolating the concentration from the standard curve.

In the second part, you will need to make a standard curve. You will prepare a series of test tubes containing increasing concentrations of nitrophenol. You will measure the absorbance of the different concentrations under alkaline condition. The absorbance will vary linearly with concentration. If you plot the concentrations of nitrophenol versus absorbance, you can use this curve to determine an unknown concentration.

MATERIALS

Equipment

- ✓ stopwatch
- ✓ 10-mL graduated cylinder
- ✓ small and large test tubes
- ✓ test-tube rack
- ✓ spectrophotometer
- ✓ small beaker
- ✓ 1-mL transfer pipet

Chemicals

- ✓ 1 mM nitrophenyl phosphate in 0.09 M citrate buffer, pH 4.8
- ✓ 0.3 M potassium hydroxide solution
- ✓ 8 mM nitrophenol standard
- ✓ Wheat germ acid phosphatase, 0.08 units/mL

PROCEDURE

Enzyme assay

1. Prepare a rack of 10 small test tubes that will fit in the spectrophotometer. Using a waterproof pen, number the tubes from 1 to 10.

2. Use a 1-mL transfer pipet to place 1 mL of KOH into each of the 10 tubes.

3. Obtain approximately 15 mL of phosphatase substrate (nitrophenyl phosphate solution) in a small beaker. Using a transfer pipet, place 1 mL of substrate solution in test tube 1. This solution will be used to determine the zero-time value of the reaction.

4. Transfer 10 mL of the phosphatase substrate from the beaker into a large test tube.

5. Using a 1-mL transfer pipet, place 0.5 mL (500 µL) of wheat germ acid phosphatase into the large test tube. Mix the tubes by shaking gently. Begin timing!

6. At the times indicated below, remove 1 mL of the solution from the large test tube. Place the solution into the tube corresponding to the time of withdrawal.

Time interval	Test tube
0 minutes	1
1 minute	2
2 minutes	3
3 minutes	4
4 minutes	5
5 minutes	6
7.5 minutes	7
10 minutes	8
15 minutes	9
20 minutes	10

7. Add 2 mL of distilled water to each of the test tubes.

Important: Do not forget this step!

Preparation of the nitrophenol standards, a serial dilution

1. Number a set of test tubes from 1 to 6.

2. Add 2 mL of water to test tubes 1 through 5 and 3.8 mL of water to test tube 6.

3. Add 0.2 mL of nitrophenol standard (8 mM) to test tube 6 and mix the contents.

4. Using your graduated cylinder, transfer 2 mL of the solution from test tube 6 to test tube 5 and mix the contents.

CAUTION!
Be sure to rinse your graduated cylinder between each dilution.

5. Transfer 2 mL of the solution from test tube 5 to test tube 4 and mix the contents. Continue in this manner until you have transferred 2 mL of the solution from test tube 3 to test tube 2 and mixed the contents.

6. Remove 2 mL of the solution from test tube 2 and discard.

7. Add 2 mL of 0.3 M KOH to each of the six test tubes.

8. Number a corresponding set of test tubes from 1 to 6. Using a transfer pipet, place 2 mL of each of these standards into a corresponding tube.

Before the addition of base, the test tubes contained the following concentrations. Use these concentrations to plot the standard curve.

Test tube	Concentration
1	0 nmol nitrophenol/mL
2	25 nmol nitrophenol/mL
3	50 nmol nitrophenol/mL
4	100 nmol nitrophenol/mL
5	200 nmol nitrophenol/mL
6	400 nmol nitrophenol/mL

Standard curve and concentration measurement

1. Allow the spectrophotometer to warm up at least 5 minutes prior to taking the readings.

2. Set the wavelength to 410 nm. You will read the standard curve first. Use test tube 1 of the standard as the blank (zero absorbance). Adjust your spectrophotometer as instructed in lab.

3. Read and record the absorbance of each of the nitrophenol standards, test tube 2 through 6.

4. Read and record the absorbance (A) of each time point in your enzyme reaction assay (test tube numbers 1 through 10).

RESULTS

1. Record the absorbance for the standard curve.

Test tube	Concentration	Absorbance
1	0 nmol nitrophenol/mL	
2	25 nmol nitrophenol/mL	
3	50 nmol nitrophenol/mL	
4	100 nmol nitrophenol/mL	
5	200 nmol nitrophenol/mL	
6	400 nmol nitrophenol/mL	

2. Prepare the standard curve for the assay. Plot the absorbance for each standard tube numbers on the y-axis as a function of the concentration of nitrophenol in the standards on the x-axis.

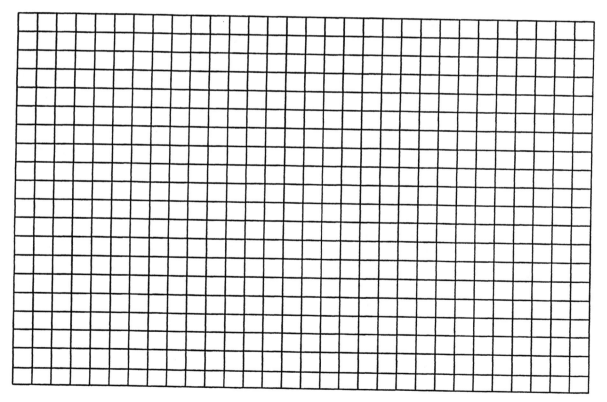

3. Using this standard curve, determine the nanomoles of nitrophenol produced in each tube of your enzyme assay.

Time (min)	Test tube	Absorbance	nitrophenol produced (nmol)
0	1	0	0
1	2		
2	3		
3	4		
4	5		
5	6		
7.5	7		
10	8		
15	9		
20	10		

1. Plot the nanomoles of nitrophenol (y-axis) produced against time (x-axis).

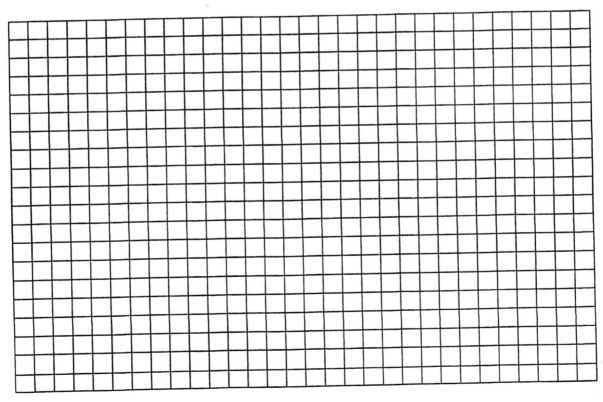

4. From the above graph, calculate the initial velocity (v_0) of the enzyme reaction in nanomoles nitrophenol produced/minute. Draw a tangent to the initial portion of the curve, where it is increasing steeply. The slope of this line is the initial velocity.

$$\text{Rate} = \text{slope} = \frac{\text{change in y}}{\text{change in x}}$$

Experiment 36

Electrophoresis of DNA and DNA fragments

This is a two-part lab. In the first lab, you will produce DNA fragments. In the second lab, you will separate the fragments by electrophoresis.

Restriction enzymes or endonucleases are enzymes that makes double-stranded cuts in DNA molecules at a particular sequence of base pairs called **recognition sequences**. The recognition sequence can be as short as four base pairs and as long as eight base pairs. Listed below are the recognition sequences for *Eco*RI and *Hin*dIII, the two enzymes that will be used in this experiment. The arrows indicate the point of cleavage within a recognition sequence. Notice the symmetry in the sequence. Both of these enzymes make a staggered four-base cut.

> Endonucleases often cleave at palindromic sites. A palindrome is word, phrase, or sentence, that is the same whether read backward or forward.

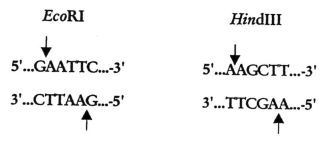

When *Eco*RI encounters this sequence in a strand of DNA, it cleaves the sugar-phosphate backbone. For example, as shown above, *Eco*RI would cleave between the G and the A on both strands of the DNA. All restriction enzymes cut DNA in a predictable and reproducible manner. Many hundreds of restriction enzymes have been discovered and catalogued according to their recognition sites. Thus, it is possible to choose from a "library" of these enzymes to cut DNA at targeted sequences.

Restriction enzymes are produced by and derived from various bacteria. Bacteria use the enzymes to destroy the DNA of infecting bacteriophages (viruses that infect bacteria). In the course of genetic investigations of bacteria, researchers noted that bacteriophages that grew well in one strain of a bacterial species frequently grew poorly (had *restricted* growth) in another strain of the same species. The growth-restricting host cells contain cleavage enzymes (restriction nucleases) that produce double-stranded breaks at a specific sequence in the phages DNA. The corresponding DNA sequence in the host cell is not attacked because the bases in the sequence are methylated. Thus, the cleavage enzymes degrade DNA from foreign sources but not from the host cell.

The DNA that will be used in this experiment is from lambda (λ), a bacteriophage that infects *Escherichia coli*. The viral DNA is a linear, double-stranded molecule of 48,502 base pairs. Restriction enzymes are named in the following way:

- *E* → First letter of genus name of host bacterium
- *co* → First two letters of species name of host bacterium
- *R* → A particular strain of this bacterial species
- *I* → The particular enzyme among several produced by this bacterial strain

The activity of these enzymes is defined in relative units (U). One unit of activity is the amount of enzyme required to digest 1 µg of λ DNA to completion in one hour in the preferred enzyme buffer at the optimal temperature for the enzyme.

The smaller DNA molecules produced by cutting DNA with one or more restriction enzymes are called **restriction fragments**. The length of a particular restriction fragment is determined by the number of base pairs that lie between the two adjacent restriction (cleavage) sites. Fragment lengths can be determined most easily by analyzing the products of a restriction enzyme cleavage reaction (digest) by electrophoresis on agarose gels. Once fragment sizes are known from the digests of several different restriction enzymes, a "map" of the cleavage sites can be constructed. The restriction site and the fragments that are generated after cleavage have no real genetic significance but are useful landmarks for creating physical maps of DNA molecules. In this two-part lab, you will cleave phage DNA and then examine the fragments produced.

Electrophoresis separates molecules based on two properties: charge and size. Nucleic acids are negatively charged at neutral pH because of the presence of phosphate groups. Because each fragment will be negatively charged, size is of the critical factor in the separation of DNA fragments. When these negatively charged molecules are placed in an electric field between two electrodes, they all migrate toward the positive electrode. In nucleic acids, each nucleotide residue contributes a negative charge from the phosphate to the overall charge of the fragment, but the mass of the nucleic acid increases correspondingly. Thus, the ratio of charge to mass remains approximately the same regardless of the size of the molecule in question. Consequently, the separation takes place simply on the basis of size and is due to the sieving action of the gel. In a given amount of time, the smaller fragments move farther than the larger ones in an electrophoretic separation.

MATERIALS

Equipment
- microfuge tubes
- water bath
- thermometer
- pipettors, volume rang 2 to 20 µL
- 10-mL graduated cylinder
- stirring rod
- 250-mL Erlenmeyer flask
- DC power supply
- gel box with leads
- watchglass

Chemicals
- λ DNA (conc. = 0.5 µg/µL)
- EcoRI (conc. = 10 units/µL)
- HindIII (conc. = 10 units/µL)
- 10× REact 2 buffer (50 mM Tris-HCl, 10 mL $MgCl_2$, 50 mM NaCl, pH 8.0)
- 1× STE buffer (100 mM NaCl, 0.04 M Tris acetate, 0.001 M EDTA)
- Sample loading buffer (30% glycerol, 50 mM EDTA, 0.25% xylene cyanol FF, 0.25% bromphenol blue)
- 50× TAE buffer (2 M Tris base, 1 M acetic acid, 0.05 M EDTA)
- agarose powder
- Carolina Blu DNA gel stain

PROCEDURE

Part A: Enzyme Digestion of DNA

1. Prepare the following samples in microfuge tubes.

Tube	water	buffer 10× REact 2	DNA	EcoRI	HindIII
1	14 µL	2 µL	4 µL λ DNA		
2	12 µL	2 µL	4 µL λ DNA	2 µL	
3	12 µL	2 µL	4 µL λ DNA		2 µL
4	10 µL	2 µL	4 µL λ DNA	2 µL	2 µL

2. Incubate the tubes for 30 minutes at 37°C.

3. Add 3 µL of sample loading buffer to all four microfuge tubes.

4. Samples should be saved for electrophoresis in the next part.

Part B: Electrophoresis of the DNA

1. Prepare 200 mL of 1× TAE buffer by taking 4 mL of 50× buffer and diluting to 200 mL with distilled H_2O.

2. To set up the gel box, place the dams at each end of the gel deck.

3. Weigh out 0.5 g of agarose and place in a beaker. Add 50 mL of 1× TAE buffer. Cover your beaker with a watchglass and heat this mixture in a microwave for 2 minutes until all the agarose particles are dissolved. Use the thermal gloves to remove the beaker from the microwave. **Caution: HOT!!**

   CAUTION! Beaker and agarose suspension will be hot. Use thermal gloves.

4. Cool the solution to about 60°C; add 2 drop of Caroline Blu DNA stain to the solution. Swirl the solution to mix in the dye.

5. Pour approximately 25 mL of slightly cooled agarose solution into the gel tray (just fill the tray up to the edge). Check to make sure that there are no bubbles in the molten agarose. Bubbles must be removed before the gel solidifies. Insert the comb at the negative electrode (black) end. Let the gel harden for 10 minutes.

6. Add 1× TAE buffer to cover the gel by about 1.5 mm (about 125 mL). Gently remove the comb and dams. Put the gel box near the power supply before loading the samples.

7. Load 15 µL of your samples into the wells. You may need to tap the tubes on the desktop to bring down any drops on the sides of the tube.

8. When all samples are loaded, close the lid on the gel box. Check that the power supply is turned off and the voltage is turned all the way down. Attach the electrodes to the gel box and power supply, making sure that the red leads connect the + terminals and the black leads connects the − terminals.

9. Turn on the power supply and adjust the voltage to 100 V.

10. Continue the electrophoresis for 45 to 60 minutes or until the purple tracking dye has moved at least 2/3 of the way down the gel.

11. Before removing the gel, turn off the power supply and detach the electric leads. Carefully transfer the gel to a staining dish. Stain the gel with Carolina Blu for 20 minutes.

12. Rinse the gel twice by it covering with water and then pouring the water off. Copy the banding pattern from the gel.

RESULTS

Draw the resulting bands on the diagram below:

Wells

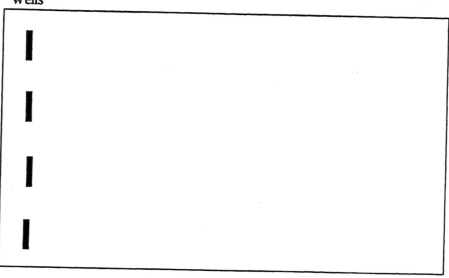

1. Sample 4 contained both *Hin*dIII and *Eco*RI. Do you expect a greater or lesser number of bands than seen when only one enzyme is used? Explain.

2. Linear DNA fragments migrate at rates inversely proportional to the log of their molecular weights. Which fragments migrate the farthest, small or large fragments? Explain.

Appendix

Hints on Graphing

A graph is used to inform the reader of the relationship between an experimentally controlled parameter (**the independent variable**) and the calculated or measured variable (**the dependent variable**). So the first thing you must do is determine which is which. For example, if you are examining the effect of temperature on the volume of a gas, you will see that as the temperature increases, the volume of the gas increases. Therefore the volume is dependent on the temperature. When plotting information on a graph, the independent variable should be plotted along the horizontal (x) axis and the dependent variable should be plotted along the vertical (y) axis. Refer to the example below as you read this section.

Choosing the appropriate scales for the dependent and independent variables causes the most difficulty for students. Remember that the goal of a graph is to present the reader with as much information as possible and to do it in a simple, clear manner. The graph itself should encompass at least one-half of the graph paper. This requirement generally means that the intersection of the x- and y-axes is not 0,0. The smallest divisions of both axes should be divisible by 2, 5, or 10, for ease of plotting and interpolation. The axis must be labeled with a symbol or word that represents the variable. Appropriate units follow the label: for example "Volume (L)."

Relationship between volume and temperature for a gas

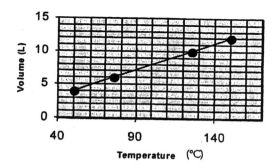

The graph consists of a line and a set of points. The line may be straight or curved, but in either case it should be smooth, not drawn from point to point. Each point should consist of a small, precise dot surrounded by a simple geometric figure, usually a circle. If more than one set of points is plotted on the same graph, squares and triangles, both filled and unfilled, may be used to distinguish between the curves. Include a legend stating what each set of data represents.

In addition to showing the relationship between two variables, a graph can be used to determine an unknown value. The graph above records four temperature/volume readings. You should be able to read the volume of the gas at any temperature between the two extremes of your measurements. This graph is called a **standard curve**. The determination of unknown values is most accurate when the readings are made within the span of the measurements of the standard curve. If you extrapolate the curve beyond the measurements, there is a chance that you are introducing error into the determination of the unknown value.